Lecture Notes in Mathematics

Edited by A. Dold and B. Eckmann

454

Joram Hirschfeld
William H. Wheeler

Forcing, Arithmetic, Division Rings

Springer-Verlag
Berlin · Heidelberg · New York 1975

Authors
Dr. Joram Hirschfeld
Department of Mathematics
Tel Aviv University
Ramat Aviv
Tel Aviv
Israel

Dr. William H. Wheeler
Bedford College
University of London
London
England

Permanent address:
Department of Mathematics
Indiana University
Bloomington, Indiana 47401
USA

Library of Congress Cataloging in Publication Data

Hirschfeld, Joram.
Forcing, arithmetic, and division rings.

(Lecture notes in mathematics ; 454)
Bibliography: p.
Includes index.
1. Forcing (Model theory) 2. Model theory.
3. Division rings. I. Wheeler, William H., 1946-
joint author. II. Title. III. Series: Lecture notes
in mathematics (Berlin) ; 454.
QA3.L28 no. 454 [QA9.7] 510'.8 [511'.8] 75-12981

AMS Subject Classifications (1970): 02H05, 02H13, 02H15, 02H20, 08A20, 10N10, 10N15, 16A40

ISBN 3-540-07157-1 Springer-Verlag Berlin · Heidelberg · New York
ISBN 0-387-07157-1 Springer-Verlag New York · Heidelberg · Berlin

Offsetdruck: Julius Beltz, Hemsbach/Bergstr.

In memory of

Abraham Robinson

ACKNOWLEDGEMENTS

We gratefully acknowledge the contributions of our colleagues and coworkers to this volume. Deserving of special mention are Mrs. S. Mandel for typing Part Two; L. Manevitz, D. Saracino, P. van Praag, and P. M. Cohn for their valuable comments and criticism of the manuscript; the members of the 1973-1974 Bedford College (University of London) logic seminar, to whom much of the material herein was presented, for their questions and comments; and the members of the Department of Mathematics of Yale University, 1969-1972, especially N. Jacobson, J. Barwise, G. Sacks, E. Fisher, M. Lerman, J. Schmerl, and S. Simpson, for their instruction and suggestions. We are indebted especially to Abraham Robinson, our adviser, for his guidance and encouragement and for the proposal of this volume. Finally, we thank our wives for their patience and moral support during the writing and preparation of this volume; the second author acknowledges in particular his gratitude to his wife for her labor of love in typing Parts One and Three.

Joram Hirschfeld

William H. Wheeler

CONTENTS

INTRODUCTION

Forcing in model theory is a recent development in the metamathematics of algebra. The context of this development has three principal features: the importance of algebraically closed fields in commutative algebra and the existence of analogues of algebraically closed fields for other algebraic systems, earlier work on model-completeness and model-completions by Abraham Robinson and others, and Paul Cohen's forcing techniques in set theory.

Algebraically closed fields serve a useful function in commutative algebra, algebraic number theory, and algebraic geometry. Certain arithmetical questions can be settled conclusively in an algebraically closed field. Examples are well-known. For instance, a system of polynomials has a common zero in some extension of their coefficient field if and only if they have a common zero in the algebraic closure of their coefficient field. In algebraic number theory, the study of the prolongations of a valuation from its base field to a finite dimensional extension field reduces to the consideration of the embeddings of the extension field into the algebraic closure of the completion of the base field. A third example is the use of universal domains in algebraic geometry as the proper setting for the study of algebraic varieties over fields. In these and other instances, the existence and use of algebraically closed fields simplify the treatment of many mathematical problems.

The usefulness of algebraically closed fields has motivated the development of analogues for other algebraic systems. The best known analogue is the class of real closed ordered fields, introduced by Artin and Schreier for the solution of Hilbert's seventeenth problem on ordered fields. Another important analogue is the Henselization of a discrete, nonarchimedean valued field (see Ax & Kochen (4)). Analogues have been introduced also for other algebraic systems, including groups (W. R. Scott (101)), abelian groups (T. Szele (106)), modules (P. Eklof & G. Sabbagh (34)), commutative rings (G. Cherlin (17)), commutative rings without nilpotent elements (D. Saracino & L. Lipshitz (98), A. Carson (12)), and on a more abstract level, universal algebra (P. M. Cohn (20), B. Jonsson (49)).

While some analogues have been quite productive, others have presented almost insuperable difficulties. Algebraically closed groups are an example of the latter. In 1951 W. R. Scott defined a group G to be algebraically closed if each finite system of equations and inequalities of the form

$$w_1 = e, w_2 = e, \ldots, w_n = e, v_1 \neq e, v_2 \neq e, \ldots, v_m \neq e$$

(where the w_i and v_j are words in indeterminates and elements of G) with a solution in some extension of G has a solution in G itself. During the next year, B. H. Neumann (70) showed that only equations need be considered in the above definition and that every algebraically closed group is simple. No more work was done on these groups until 1969, when Neumann (71) proved that each finitely generated, recursively absolutely presented group is a subgroup of every algebraically closed group. Since algebraically closed groups are simple, they have not been amenable to the usual methods of group theory. However, they can be investigated through forcing techniques.

Algebraically closed fields have a special significance for mathematical logic also. The theory of algebraically closed fields is the canonical example of each of the following concepts: a complete theory, a model-complete theory, a totally transcendental theory, and an $\aleph_1$-categorical but not $\aleph_0$-categorical theory. Moreover, algebraically closed fields were the starting point for the metamathematics of algebra. Relationships between the theory of fields and the theory of algebraically closed fields led to the concepts of model-completeness and model-completion (A. Robinson).

From the point of view of algebra, the concept of a model-completion has been the most useful metamathematical concept introduced by logicians. For example, algebraically closed fields are the model-completion of commutative fields; real closed ordered fields are the model-completion of ordered fields; Hensel fields are the model-completion of discrete, nonarchimedean valued fields; and Szele's algebraically closed abelian groups are the model-completion of abelian groups. Both the Nullstellensatz and Hilbert's seventeenth problem are consequences of the model-completeness of algebraically closed fields and of real closed ordered fields, respectively. Furthermore, the notion of a model-completion led A. Robinson to the definition and proof of the existence of differentially closed fields.

The notion of a model-completion was weakened in 1969 by Eli Bers to that of a model-companion. A theory T^* is called the model-companion of a theory T if (i) T and T^* are mutually model-consistent, i.e., each model of T is contained in a model of T^* and vice versa, and (ii) T^* is model-complete, i.e., whenever M and M' are models of T^* and M' contains M, then any sentence defined in M is true in M if and only if it is true in M'. Any model-completion is also a model-companion. However, some theories, for example, formally real fields, have a model-companion but not a model-completion. This phenomenon occurs when the original theory does not have the amalgamation property. The concept of a model-companion encompasses all of the common, useful analogues in algebra of algebraically closed fields.

The second requirement in the definition of a model-companion, that T^* must be model-complete, is the essence of "being algebraically closed" and is also the more difficult of the two requirements to satisfy. Robinson's model-completeness test demonstrates the first assertion. A formula $\phi(v_0,\ldots, v_n)$ in a first order logic is called primitive if it consists of a string of existential quantifiers followed by a conjunction of atomic and negated atomic formulas. Robinson's model-completeness test states that a theory T^* is model-complete if and only if whenever M and M' are models of T^* with M contained in M' and $\phi(a_0,\ldots, a_n)$ is a primitive sentence defined in M, then M satisfies $\phi(a_0,\ldots, a_n)$ if and only if M' satisfies $\phi(a_0,\ldots, a_n)$. For fields, an equivalent form of this test is that every finite system of polynomials with coefficients from M has a solution in M if and only if it has a solution in M'. The difficulty of achieving the second requirement for a model-companion is illustrated by the case of groups. The collection of algebraically closed groups is precisely the class of groups which satisfies the latter condition of Robinson's model-completeness test. But the class of algebraically closed groups does not give rise to a model-companion for the class of groups, because the class of algebraically closed groups cannot be axiomatized (Eklof & Sabbagh (34)). This nonaxiomatizability is due to the impossibility of determining algorithmically which sets of equations should have solutions. Cases such as groups require the generalizations of a model-companion introduced as part of forcing in model theory.

Forcing in model theory was motivated by Paul Cohen's forcing techniques in set theory. Forcing in set theory is a method for the gradual construction of a new model of set theory from an old model. The construction requires a countably infinite number of steps. At each step in the construction only a finite number of the new membership relations have been decided. Each of these finite amounts of information constitutes a condition. Every property of the new model is determined by one of these conditions.

Abraham Robinson adapted Cohen's forcing techniques from set theory to model theory during 1969 and 1970. Robinson developed two types of forcing for model theory - finite forcing and infinite forcing. Application of either type of forcing relative to a first order theory constructs generic models and a companion theory for the original theory. Forcing in model theory provided unexpectedly a new perspective on model-completions and model-companions. For example, application of forcing to commutative fields yields the algebraically closed fields; for formally real fields, the result is the real closed fields. Indeed, whenever a theory has either a model-completion or a model-companion, then application of forcing yields that model-completion or model-companion. Forcing also constructs analogues of algebraically closed fields for theories without model-companions. For example, the generic groups constructed by either finite or infinite forcing are algebraically closed in the sense of Scott. Moreover, finite forcing enabled one for the first time to construct algebraically closed groups while restricting the finitely generated subgroups. Angus Macintyre (56) used finite forcing to prove the converse to Neumann's theorem, mentioned above, on finitely generated subgroups of algebraically closed groups. Thus, forcing in model theory provides a uniform method for obtaining and investigating analogues of algebraically closed fields for other algebraic systems.

In this report, we begin with a discussion of forcing and the various metamathematical analogues of algebraically closed fields. Then we investigate these analogues for models of arithmetic and for division rings.

Forcing and related concepts are discussed in Part One. The topics are existentially complete structures, existentially universal structures, model-completions and model-companions, infinite forcing, approximating chains for the class of infinitely generic structures, finite forcing, axiomatization results, and recursive aspects of forcing.

Existentially complete and generic structures for arithmetic are examined in Part Two by J. Hirschfeld. The countable, existentially complete models are analyzed in considerable depth. Preliminary results on partial recursive functions lead to the introduction of existential closures of elements. The existential closure of a single element is called a simple model. The simple models coincide with the r.e. ultrapowers of the natural numbers. Each countable model of arithmetic can be embedded in one of these simple models. Next, a regular model is defined as an existentially complete model with no cofinal simple submodels. One of the results on regular models is the existence of a correspondence between regular models and structures for second order arithmetic. In particular, the infinitely generic models can be identified by their associated structures for second order arithmetic. Moreover, this correspondence relates an approximating chain for the infinitely generic models to the β_n models of second order arithmetic.

Existentially complete and generic division rings are examined in Part Three by W. Wheeler. Arithmetical definitions of various higher order concepts, such as transcendentalness, are formulated for existentially complete division rings. These definitions are used in answering algebraic questions on subfields, centralizers, and embeddings. For instance, the maximal subfields of a countable, existentially complete division ring are described, and a proper endomorphism of each such division ring is constructed. Next, a Nullstellensatz for noncommutative polynomials is proven. Lastly, the forcing classes of division rings are investigated through structures for second order arithmetic which are arithmetically defined within existentially complete division rings.

CONVENTIONS AND PRELIMINARIES

We will use the customary conventions of mathematical logic (see Robinson (79), Shoenfield (103)). Explicitly, we will always be using a first order language $\mathcal{L}$ with logical symbols $\wedge$, $\vee$, $\sim$, $\exists$, and $=$. The universal quantifier $\forall$ will be regarded as an abbreviation for $\sim\exists\sim$. The variables in $\mathcal{L}$ will be $v_0, v_1, v_2, \ldots$; however, the symbols x, y, z may sometimes be used in place of the actual variables. The language may also have appropriate nonlogical symbols. For models of arithmetic, the language will include the binary function symbols $+$ and $\cdot$, the binary relation symbol $<$, and two constant symbols 0 and 1. For division rings, the language will include the binary function symbols $+$ and $\cdot$ and the constant symbols 0 and 1. Syntactic and semantic concepts for $\mathcal{L}$ are defined as usual.

Concerning notation, formulas will usually be denoted by Greek letters, either lower or upper case. If ϕ is a formula, then the notation $\phi(v_0, \ldots, v_n)$ will mean that the free variables of ϕ are among $v_0, \ldots, v_n$. First order structures will be denoted by upper case Roman letters. Models for second order arithmetic will be denoted by upper case Fraktur. If ϕ is a formula of $\mathcal{L}$ with free variables among $v_0, \ldots, v_n$ and if $a_0, \ldots, a_n$ are constant symbols naming elements of a structure M for $\mathcal{L}$, then the formula $\phi(a_0, \ldots, a_n)$ is said to be <u>defined in M</u>. If $\phi(a_0, \ldots, a_n)$ is defined in M, then "M satisfies $\phi(a_0, \ldots, a_n)$" will be denoted by $M \models \phi(a_0, \ldots, a_n)$.

The distinction between elements of a structure and names for these elements in some language will be blurred deliberately. If M is a structure for $\mathcal{L}$, then the new language obtained by adding names for each element of M to $\mathcal{L}$ will be called the language of M and will be denoted by $\mathcal{L}(M)$. The (basic) diagram of a structure M, denoted by Diag(M), is the set of all atomic and negated atomic sentences in $\mathcal{L}(M)$ which are true in M.

A formula is in <u>prenex normal form</u> if it consists of a string of quantifiers followed by a quantifier-free formula. Every formula is logically equivalent to a formula in prenex normal form. A <u>block of quantifiers</u> in a formula in prenex normal form is a sequence of like quantifiers which is neither immediately preceded nor immediately followed by another quantifier of the same type. A formula in prenex normal form is called an <u>$\exists_n$ formula</u> if either it has fewer than n blocks of quantifiers or it has exactly n blocks of quantifiers beginning at the left with a block of existential quantifiers. The <u>$\forall_n$ formulas</u> are defined similarly. A formula of $\mathcal{L}$ is called <u>existential</u> if either it is an atomic formula or the negation of an atomic formula or it is formed from atomic and negated atomic formulas by successive conjunctions, disjunctions, or existential quantifications. <u>Universal formulas</u> are defined similarly. A formula is a <u>universal existential formula</u> if either it is an existential formula or it is formed from existential formulas through successive conjunctions, disjunctions, and universal quantifications. Clearly, every existential formula is logically equivalent to an $\exists_1$ formula, and so forth.

A <u>theory</u> T in $\mathcal{L}$ is a set of consistent sentences of $\mathcal{L}$. If ϕ is a sentence of $\mathcal{L}$, then $T \vdash \phi$ will denote that ϕ is deducible from T. The set of all $\forall_n$ formulas deducible from a theory T will be denoted by $T_{\forall_n}$. The set $T_{\exists_n}$ of sentences is defined similarly. For small values of n, the following notation is more convenient: $T_\exists$ for $T_{\exists_1}$, $T_\forall$ for $T_{\forall_1}$, $T_{\forall\exists}$ for $T_{\forall_2}$, and so forth.

The class of models of a theory T will be denoted by $\mathcal{M}(T)$. If Σ is a class of structures for $\mathcal{L}$, then $\mathcal{T}h(\Sigma)$ will denote the theory of the class Σ, i.e., the set of sentences of $\mathcal{L}$ which are true in every member of Σ. A class of structures for $\mathcal{L}$ is said to be a <u>generalized elementary class</u>, an <u>axiomatizable class</u>, or an <u>EC_Δ class</u> if there is a (possibly infinite) set T of sentences from $\mathcal{L}$ such that $\Sigma = \mathcal{M}(T)$. In particular, Σ is a generalized elementary class if and only if $\Sigma = \mathcal{M}(\mathcal{TA}(\Sigma))$. A class of structures for $\mathcal{L}$ is said to be <u>inductive</u> if whenever $\{M_\alpha : \alpha < \gamma\}$ is an ascending chain of members of the class with $M_\alpha \subseteq M_\beta$ for $\alpha < \beta$ (α, β, and γ denote ordinals), then $\bigcup_{\alpha<\gamma} M_\alpha$ is also a member of the class. A theory is <u>inductive</u> if the class of models of the theory is inductive. It is well-known that a theory is inductive if and only if it can be axiomatized by $\forall\exists$ formulas.

If M is a structure for $\mathcal{L}$, then $\mathcal{TA}(M)$ will denote the set of sentences of $\mathcal{L}$ which are true in M. The set of sentences in the language $\mathcal{L}(M)$ of M which are true in M will be denoted by $\mathcal{TA}(M, \bar{m})$.

The infinitary languages $\mathcal{L}_{\kappa,\omega}$ and $\mathcal{L}_{\infty,\omega}$ will be used occasionally (Karp (50, 51), Keisler (52)).

Two structures M and M' for the language $\mathcal{L}$ are said to be <u>elementarily equivalent</u> if M and M' satisfy exactly the same sentences sentences of $\mathcal{L}$. That M and M' are elementarily equivalent will be denoted by $M \equiv M'$. If M is contained in M', then M is said to be an <u>elementary substructure</u> of M' (M' is an <u>elementary extension</u> of M) if whenever $\phi(a_0,\ldots, a_n)$ is a sentence defined in M, then M satisfies $\phi(a_0,\ldots, a_n)$ if and only if M' satisfies $\phi(a_0,\ldots, a_n)$. That M is an elementary substructure of M' will be denoted by $M \prec M'$. The relationships of $\mathcal{L}_{\kappa,\omega}$-<u>equivalence</u>, $\mathcal{L}_{\infty,\omega}$-<u>equivalence</u>, $\mathcal{L}_{\kappa,\omega}$-<u>substructure</u>, and $\mathcal{L}_{\infty,\omega}$-<u>substructure</u> for structures M and M' are defined similarly. These relationships are denoted symbolically by $M \equiv_{\kappa,\omega} M'$, $M \equiv_{\infty,\omega} M'$, $M \prec_{\kappa,\omega} M'$, and $M \prec_{\infty,\omega} M'$, respectively.

The Deduction Theorem, the Completeness Theorem, and the Compactness Theorem for first order logic will be used implicitly. For the sake of reference, these are as follows:

<u>Deduction Theorem</u>. A set of sentences is consistent if and only if each of its finite subsets is consistent.

Completeness Theorem. A set of sentences is consistent if and only if it has a model.

Compactness Theorem. A set of sentences has a model if and only if each of its finite subsets has a model.

PART ONE

FORCING

Part One is a comprehensive, although nonexhaustive, discussion of forcing in model theory and of the various, metamathematically defined analogues for algebraically closed fields. The existentially complete structures are investigated first; these are the most direct and inclusive analogue of algebraically closed fields. Next, the existentially universal structures, analogues of universal domains, are developed along parallels with the existentially complete structures. The existentially universal structures form a subclass of the class of existentially complete structures. Of the other classes which will be studied, most lie between these two classes.

Some formal notions are necessary as background for the introduction of the forcing classes. Three types of companion theories for a first order theory are discussed. The first of these is the model-completion of a theory. When a model-completion exists for a theory, its role is exactly analogous to the role of the theory of algebraically closed fields. Unfortunately, a model-completion does not exist in many instances. This situation leads to the introduction of the model-companion and the forcing companions of a theory.

Forcing in model theory, although motivated by forcing in set theory, in fact generalizes the notion of a model-companion. Infinite forcing is developed in both its formal and material aspects. Some important results are the reduction theorem, the characterization of the class of infinitely generic structures, and the approximating chain constructions for the class of infinitely generic structures. After infinite forcing, finite forcing and the class of finitely generic structures are discussed.

Next, axiomatizability results for the various analogues, including Saracino's theorem on the existence of model-companions for $\aleph_0$-categorical theories, are presented. Then the recursive aspects of the forcing constructions are investigated by encoding the constructions within the arithmetical and analytical hierarchies of sets. Part One concludes with a brief review of the basic results and a summary of the current state of knowledge about the various analogues for specific algebraic systems.

The results in Part One are due to many individuals. Rather than giving credit for each result separately, we shall indicate some references for each chapter:

Chapter	References
1	Eklof & Sabbagh (34), Simmons (104), Robinson (81, 82), Wood (109)
2	Robinson (79, 80, 81)
3	Robinson (81, 82)
4	Cherlin (14, 15), Hirschfeld (46) Saracino (93), Wheeler (107)
5	Robinson (80), Barwise & Robinson (6); for applications of finite forcing, see Macintyre (56, 57, 58)
6	Saracino (94, 95), Wood (108, 109)
7	Hirschfeld (46), Wheeler (107).

CHAPTER 1

EXISTENTIALLY COMPLETE STRUCTURES AND EXISTENTIALLY UNIVERSAL STRUCTURES

§ 1 Existentially Complete Structures

The existentially complete structures are the most direct analogue of algebraically closed fields. Recall that a field K is algebraically closed if every polynomial in one variable with coefficients from K has a zero in K. This is equivalent via Hilbert's Nullstellensatz to the following, apparently stronger condition: if $p_1(x_1,\ldots, x_n), \ldots, p_r(x_1,\ldots, x_n), q_1(x_1,\ldots, x_n), \ldots, q_s(x_1,\ldots, x_n)$ are polynomials with coefficients from K such that in some extension L of K there are elements $a_1', \ldots, a_n'$ satisfying

$$p_1(a_1',\ldots, a_n') = 0, \ldots, p_r(a_1',\ldots, a_n') = 0,$$
$$q_1(a_1',\ldots, a_n') \neq 0, \ldots, q_s(a_1',\ldots, a_n') \neq 0,$$

then there are elements $a_1, \ldots, a_n$ of K such that

$$p_1(a_1,\ldots, a_n) = 0, \ldots, p_r(a_1,\ldots, a_n) = 0$$
$$q_1(a_1,\ldots, a_n) \neq 0,\ldots, q_s(a_1,\ldots, a_n) \neq 0.$$

The statement that there exist $a_1, \ldots, a_n$ which are simultaneously a zero of the polynomials p_i and a nonzero of the polynomials q_j can be formulated as an existential sentence with constants from K in the first order language for fields. Conversely, every existential sentence with constants from K in the first order language for fields is equivalent in the theory of fields to a statement that, for some fixed collection of polynomials $p_1, \ldots, p_r, q_1, \ldots, q_s$ with coefficients from K, there are elements of K which are simultaneously a zero of $p_1, \ldots, p_r$ and a nonzero of $q_1, \ldots, q_s$. So, existential sentences about fields and statements about the existence of simultaneous zeroes and nonzeroes for two systems of polynomials are equivalent to one another. Thus, a field K is algebraically closed if and only if whenever $\phi(v_0,\ldots, v_m)$ is an existential formula in the language of fields and $b_0, \ldots, b_m$ are elements of K, then $\phi(b_0,\ldots, b_m)$ is true in K if and only if it is true in some extension of K. An algebraically closed field is "existentially complete" in the sense that every existential sentence defined in it and true in some extension of it is already true in it.

A substructure M of a structure M' is said to be existentially complete in M' if every existential sentence which is defined in M and true in M' is true in M also. Equivalently, a substructure M of a structure M' is existentially complete in M' if and only if every universal sentence which is defined in M and true in M is true in M'. A structure M in a class Σ of similar structures is said to be existentially complete in Σ if M is existentially complete in every extension M' of M in Σ. When the class Σ is fixed, the clarifying phrase "in Σ" is omitted. The class of existentially complete members of Σ will be denoted by $\mathcal{E}_\Sigma$.

For example, if Σ is the class of fields, then $\mathcal{E}_\Sigma$ is the class of algebraically closed fields. When Σ is the class of formally real fields, then $\mathcal{E}_\Sigma$ is the class of real closed fields. For the class of discrete, nonarchimedean valued fields, $\mathcal{E}_\Sigma$ is the class of Hensel fields. If Σ is the class of groups or the class of abelian groups, then $\mathcal{E}_\Sigma$ is the class of algebraically closed groups or the class of algebraically closed abelian groups, respectively. Finally, if Σ is the class of totally ordered sets, then $\mathcal{E}_\Sigma$ is the class of densely totally ordered sets.

An arbitrary class of structures may have no existentially complete members at all. For example, the set $\{Q(\sqrt[n]{2}) : n < \omega\}$ of fields has no existentially complete members. Each member of this class is contained in another member which has a higher order root of 2.

In order to ensure the existence of existentially complete structures, one usually requires that the class of structures be inductive.

The models of an arbitrary theory need not form an inductive class. Consequently, each model of the theory may fail to be existentially complete in the class of models of the theory. For example, consider a set of axioms for totally ordered sets with endpoints in the language with a binary relation $<$ and no constant symbols. Every model of this theory is contained in a larger model with new endpoints, so no model can be existentially complete. However, there is a natural class of totally ordered sets, each of which is contained in a totally ordered set with endpoints and is existentially complete in every such extension - namely, the class of densely, totally ordered sets. Reflection on this example indicates that a structure M should be called existentially complete for a theory T if (i) M is a structure for the language of T, (ii) M is contained in a model of T, and (iii) M is existentially complete in every model of T which extends M. The class of existentially complete models for a theory T will be denoted by $\mathcal{E}_T$, or merely by $\mathcal{E}$ if the theory T is clear from context.

This definition of existentially complete structures for a theory is related to the previous definition for a class as follows. A structure is contained in a model of a theory T if and only if it is a model of $T_\forall$, so the class $\Sigma_T = \mathcal{M}(T_\forall)$ consists of all substructures of models of T. Every existentially complete structure for T is in Σ_T. Moreover, a structure in Σ_T is existentially complete in Σ_T if and only if it is existentially complete for the theory T. Hence $\mathcal{E}_{\mathcal{M}(T_\forall)} = \mathcal{E}_T$. For example, when one considers the theory T of fields in a language with no function symbol for multiplicative inverse, the class Σ_T is the class of integral domains. Nevertheless, $\mathcal{E}_{\Sigma_T}$ is still the class of algebraically closed fields, for the assertion that an element has an inverse is an existential sentence. Thus, for a discussion of the existentially complete structures for a theory T, one need consider only the class of structures $\Sigma_T = \mathcal{M}(T_\forall)$. Since $T_\forall$ is a universal theory, the class Σ_T is an inductive class, and existentially complete structures will exist (see Proposition 1.3 below).

The following criterion for a structure to be existentially complete in an extension will be useful later.

<u>Proposition 1.1</u>. A structure M is existentially complete in an extension M' if and only if there is an elementary extension M'' of M such that the following diagram commutes:

$$\begin{array}{ccc} & \subseteq & M'' \\ M' & & \curlyvee \\ & \supseteq & M \end{array} .$$

<u>Proof</u>. Assume M is existentially complete in M'. Let T be the complete theory of M in the language $\mathcal{L}(M)$. It suffices to show that $T \cup \mathrm{Diag}(M')$ is consistent, where the same constants are used to name elements of M in both T and $\mathrm{Diag}(M')$. Let $\psi(b_0, \ldots, b_n)$ be a finite conjunction of sentences from $\mathrm{Diag}(M')$, where the constants naming elements of $M' - M$ have been displayed. The sentence $\psi(b_0, \ldots, b_n)$ is consistent with T if and only if the sentence $\exists v_0 \ldots \exists v_n\, \psi(v_0, \ldots, v_n)$ is consistent with T. Since M' satisfies $\exists v_0 \ldots \exists v_n\, \psi(v_0, \ldots, v_n)$ and M is existentially complete in M', M also satisfies $\exists v_0 \ldots \exists v_n\, \psi(v_0, \ldots, v_n)$. Thus, the sentence $\exists v_0 \ldots \exists v_n\, \psi(v_0, \ldots, v_n)$ is actually in T and so is consistent with T.

The converse is immediate.

Existential completeness was defined in terms of "pulling down" existential sentences. But once a structure is existentially complete, it can pull down slightly more.

Proposition 1.2. (1) If M is existentially complete in M' and ϕ is an $\forall\exists$ sentence which is defined in M and is true in M', then ϕ is true in M.

(2) If M is existentially complete in a class Σ and ϕ is an $\forall\exists$ sentence which is defined in M and is true in some extension of M in Σ, then ϕ is true in M.

Proof. (1) Since ϕ is an $\forall\exists$ sentence, ϕ has the form $\forall v_0 \ldots \forall v_n\, \psi(v_0,\ldots, v_n)$ where ψ is an existential formula. Let $a_0,\ldots, a_n$ be arbitrary but fixed elements of M. Then $\psi(a_0,\ldots, a_n)$ is defined in M and therefore in M' also. Since M' satisfies ϕ, M' satisfies $\psi(a_0,\ldots, a_n)$. Because M is existentially complete in M', M satisfies $\psi(a_0,\ldots, a_n)$ also. Since $a_0,\ldots, a_n$ were arbitrary elements of M, M must satisfy $\forall v_0 \ldots \forall v_n\, \psi(v_0,\ldots, v_n)$, that is, ϕ is true in M.

(2) Immediate from (1).

The basic facts about the class of existentially complete structures are divided into two groups - those which are true for any inductive class, and those which require in addition that the class is a generalized elementary class. Initially, we shall assume only inductivity.

Proposition 1.3. Assume Σ is an inductive class. Each member of Σ is contained in an existentially complete member Σ.

Proof. Let M be a member of Σ. Let $\kappa = \max\{\mathrm{card}(M), \mathrm{card}\,(\mathcal{L}), \aleph_0\}$ and let $\{\phi_\alpha : \alpha < \kappa\}$ be an enumeration of all the existential sentences in the language of M, i.e., sentences which are defined in M.

Let $M_0 = M$. Define a chain $\{M_\alpha : \alpha < \kappa\}$ of structures from Σ inductively as follow:

(i) $\alpha = \beta + 1$ is a successor ordinal. If ϕ_β is true in some extension M' of M_β in Σ, then let $M_\alpha = M'$. Otherwise, let $M_\alpha = M_\beta$.

(ii) α is a limit ordinal. Let $M_\alpha = \bigcup_{\beta<\alpha} M_\alpha$. M_α is a member of Σ, since Σ is inductive.

Now let $M^1 = \bigcup_{\alpha<\kappa} M_\alpha$. The structure M^1 is a member of Σ, since Σ is inductive; and any existential sentence defined in M_0 and true in some extension of M^1 in Σ is already true in M^1 itself.

Apply the same construction to M^1 to obtain a member M^2 of Σ with the property that any existential sentence defined in M^1 and true in some extension of M^2 in Σ is already true in M^2. Iterating this process countably many times yields a chain $M = M^0 \subseteq M^1 \subseteq M^2 \subseteq \ldots \subseteq M^n \subseteq \ldots$, $n < \omega$. Now let $M^\omega = \bigcup_{n<\omega} M^n$. Clearly, M^ω contains M. The structure M^ω is a member of Σ, since each M^n is in Σ and Σ is inductive. Moreover, M^ω is existentially complete. To verify this, let ψ be an existential sentence defined in M^ω and true in some extension of M^ω in Σ. As ψ contains names of only finitely many elements of M^ω, ψ is defined in M^r for some $r < \omega$. Since ψ is true in some extension of M^r in Σ, ψ is true in M^{r+1}. Hence, ψ is true in M^ω (if an existential sentence is true in some structure, then it is true in any extension of that structure).

The class of existentially complete structures was defined in a local maner . However, it can also be characterized globally by its own properties.

Proposition 1.4. Assume Σ is an inductive class. Then $\mathcal{E}_\Sigma$ is the unique subclass $\mathcal{C}$ of Σ satisfying

(1) every member of Σ is contained in a member of $\mathcal{C}$;

(2) if M and M' are members of $\mathcal{C}$ and M' contains M, then M is existentially complete in M';

(3) $\mathcal{C}$ contains every other subclass of Σ which satisfies conditions (1) and (2).

Proof. The class of existentially complete structures satisfies condition (2) obviously, and according to the preceding proposition it satisfies condition (1) also. To complete the proof, it suffices to show that any subclass of Σ satisfying conditions (1) and (2) is a subclass of $\mathcal{E}_\Sigma$. Suppose $\mathcal{D}$ is a subclass satisfying (1) and (2). Let M be a member of $\mathcal{D}$, and let ψ be an existential sentence defined in M and satisfied in some extension M' of M in Σ. Since $\mathcal{D}$ satisfies (1), there is a member M" of $\mathcal{D}$ which contains M'. Since ψ is existential and is true in M', ψ is true in M". Since $\mathcal{D}$ satisfies (2), ψ is true in M. Thus, M is existentially complete in Σ, and $\mathcal{D}$ is contained in $\mathcal{E}_\Sigma$.

Proposition 1.5. Assume Σ is an inductive class. Then $\mathcal{E}_\Sigma$ is inductive also.

Proof. This follows from the inductivity of Σ and the fact that true existential sentences are persistent under extension.

To go further, one must assume that Σ is a generalized elementary class. This assumption contributes two things: first, the property of existential completeness in Σ is a more intrinsic property of a structure and can be characterized syntactically; and secondly, certain new structures constructed by compactness arguments are members of Σ also. In the remainder of this section, Σ will always be a generalized elementary class. A set of axioms for Σ in an appropriate, first order language $\mathcal{L}(\Sigma)$ for Σ will be denoted by T_Σ.

Although inductivity of Σ is unnecessary for some of the results, one should assume implicitly that Σ is inductive in order to ensure that the results are not vacuously true. It is well-known that a generalized elementary class is inductive if and only if it can be axiomatized by $\forall\exists$ sentences. The paradigm of an inductive, generalized elementary class is the class $\Sigma_T = \mathcal{M}(T_\forall)$ for a first order theory T

The following proposition provides a syntactical characterization of existentially complete structures.

Proposition 1.6. Assume Σ is a generalized elementary class. The following are equivalent for a member M of Σ:

(i) M is existentially complete in Σ;

(ii) for each existential sentence ϕ defined in M, M satisfies ϕ if and only if $T_\Sigma \cup \mathrm{Diag}(M) \cup \{\phi\}$ is consistent;

(iii) for each universal formula $\psi(v_0,\ldots, v_n)$ in the language $\mathcal{L}(\Sigma)$ of Σ and for each sequence $a_0,\ldots, a_n$ of elements of M, M satisfies $\psi(a_0,\ldots, a_n)$ if and only if there is an existential formula $\phi(v_0,\ldots, v_n)$ in the language $\mathcal{L}(\Sigma)$ such that M satisfies $\phi(a_0,\ldots, a_n)$ and $T_\Sigma \vdash \forall v_0 \ldots \forall v_n\, (\phi(v_0,\ldots, v_n) \to \psi(v_0,\ldots, v_n))$.

Proof. (i) implies (ii). Assume M is existentially complete. If M satisfies an existential sentence ϕ, then clearly $T_\Sigma \cup \mathrm{Diag}(M) \cup \{\phi\}$ is a consistent set of sentences. Conversely, suppose that ϕ is an existential sentence defined in M and consistent with $T_\Sigma \cup \mathrm{Diag}(M)$ Then $T_\Sigma \cup \mathrm{Diag}(M) \cup \{\phi\}$ has a model M'. The structure M' is in Σ, since it satisfies T_Σ; and M' extends M, because M' is a model of Diag(M). Since M is existentially complete in Σ and ϕ is an existential sentence defined in M and true in the extension M' of M in Σ, ϕ is true in M.

(ii) implies (iii). Assume that M satisfies part (ii). The sufficiency of the second condition in part (iii) is clear. To prove necessity, suppose that M satisfies a universal sentence $\psi(a_0, \ldots, a_n)$. Then M does not satisfy $\sim\psi(a_0, \ldots, a_n)$, which is logically equivalent to an existential sentence. Since M satisfies part (ii), the set $T_\Sigma \cup \mathrm{Diag}(M) \cup \{\sim\psi(a_0, \ldots, a_n)\}$ must be inconsistent. Consequently, there is a conjunction $\phi'(a_0, \ldots, a_n, a_{n+1}, \ldots, a_m)$ of basic sentences from the diagram of M such that $T_\Sigma \vdash (\phi'(a_0, \ldots, a_n, a_{n+1}, \ldots, a_m) \to \psi(a_0, \ldots, a_n))$, where $a_0, \ldots, a_n, a_{n+1}, \ldots, a_m$ are all of the constant symbols occurring in either ψ or ϕ' but not in $\mathcal{L}(\Sigma)$. Then $T_\Sigma \vdash (\exists v_{n+1} \cdots \exists v_m \phi'(a_0, \ldots, a_n, v_{n+1}, \ldots, v_m) \to \psi(a_0, \ldots, a_n))$, and

$$T_\Sigma \vdash \forall v_0 \cdots \forall v_n (\exists v_{n+1} \cdots \exists v_m \phi'(v_0, \ldots, v_n, v_{n+1}, \ldots, v_m) \to \psi(v_0, \ldots, v_n)).$$

(iii) implies (i). If M satisfies part (iii), then any universal sentence defined and true in M must be true in any extension of M in Σ, so M is existentially complete in Σ.

The next series of lemmas will lead to an alternative form of the maximality condition in Proposition 1.4.

Lemma 1.7. Assume Σ is a generalized elementary class. If M and M' are members of Σ and M is existentially complete in M', then an existential sentence defined in M is consistent with $T_\Sigma \cup \mathrm{Diag}(M)$ if and only if it is consistent with $T_\Sigma \cup \mathrm{Diag}(M')$.

Proof. Since M' contains M, consistency with $T_\Sigma \cup \mathrm{Diag}(M')$ implies consistency with $T_\Sigma \cup \mathrm{Diag}(M)$.

Conversely, suppose that ϕ is an existential sentence defined in M and inconsistent with $T_\Sigma \cup \mathrm{Diag}(M')$. Then there is a finite conjunction $\psi(b_0, \ldots, b_n)$ of basic sentences from the diagram of M', where $b_0, \ldots, b_n$ are the constants which name elements of M' - M, such that $T_\Sigma \cup \text{Diag}(M) \cup \{\psi(b_0,\ldots, b_n)\} \vdash \sim\phi$. Then $T_\Sigma \cup \text{Diag}(M) \vdash \psi(b_0,\ldots, b_n) \to \sim\phi$, so $T_\Sigma \cup \text{Diag}(M) \vdash \forall v_0 \ldots \forall v_n (\psi(v_0,\ldots, v_n) \to \sim\phi)$. Since M is existentially complete in M' and M' satisfies $\exists v_0 \ldots \exists v_n \psi(v_0,\ldots, v_n)$, M satisfies this sentence also. Therefore, $T_\Sigma \cup \text{Diag}(M) \vdash \sim\phi$. Consequently, ϕ is inconsistent with $T_\Sigma \cup \text{Diag}(M)$.

Lemma 1.8. Assume Σ is a generalized elementary class. Let M, M', and M'' be members of Σ such that M' and M'' extend M and M is existentially complete in M''. Then there is a structure M''' in Σ such that

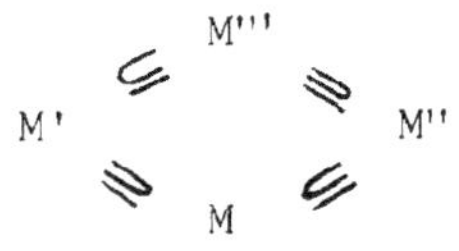

Proof. Introduce constant symbols to name each element of M' and M'' such that each element of M is given the same name in both M' and M'' and such that no element of $M' - M$ has the same name as an element of M'' and vice versa. The desired M''' exists if and only if $T_\Sigma \cup \text{Diag}(M') \cup \text{Diag}(M'')$ is consistent.

It suffices to show that any finite subset of $\text{Diag}(M'')$ is consistent with $T_\Sigma \cup \text{Diag}(M')$. Let $\psi(b_0,\ldots, b_n)$ be a conjunction of sentences from $\text{Diag}(M'')$, where $b_0,\ldots, b_n$ are the constants in ψ which name elements of $M'' - M$. Then $\psi(b_0,\ldots, b_n)$ is consistent with $T_\Sigma \cup \text{Diag}(M')$ if and only if $\exists v_0 \ldots \exists v_n \psi(v_0,\ldots, v_n)$ is consistent with $T_\Sigma \cup \text{Diag}(M')$. Now $\exists v_0 \ldots \exists v_n \psi(v_0,\ldots, v_n)$ is an existential sentence defined in M, and it is consistent with $T_\Sigma \cup \text{Diag}(M)$, since M'' contains M. According to the preceding lemma, $\exists v_0 \ldots \exists v_n \psi(v_0,\ldots, v_n)$ is consistent with $T_\Sigma \cup \text{Diag}(M')$.

Corollary 1.9. Assume Σ is a generalized elementary class. If M, M', and M'' are members of Σ such that M is existentially complete in Σ and is a substructure of both M' and M'', then there is a member M''' of Σ such that

$$\begin{array}{ccccc} & & M''' & & \\ & \subseteq & & \supseteq & \\ M' & & & & M'' \\ & \supseteq & & \subseteq & \\ & & M & & \end{array}$$

Thus, one can always amalgamate two structures if the amalgam is existentially complete.

Lemma 1.10. Assume Σ is a generalized elementary class. Suppose M and M' are members of Σ such that M' is existentially complete in Σ, M is a substructure of M', and M is existentially complete in M'. Then M is existentially complete in Σ.

Proof. Let ϕ be an existential sentence defined in M and true in some extension M'' of M in Σ. By Lemma 1.8, there is a structure M''' such that

$$\begin{array}{ccccc} & & M''' & & \\ & \subseteq & & \supseteq & \\ M' & & & & M'' \\ & \supseteq & & \subseteq & \\ & & M & & \end{array}$$

Since ϕ is existential, M''' satisfies ϕ. Since M' is existentially complete and ϕ is defined in M' and true in M''', M' satisfies ϕ. Finally, since M is existentially complete in M', M satisfies ϕ. Thus, M is existentially complete in Σ.

Proposition 1.4 can now be reformulated as follows.

Proposition 1.11. Assume that Σ is an inductive, generalized elementary class. Then $\mathcal{E}_\Sigma$ is the unique subclass $\mathcal{C}$ of Σ satisfying

(1) each member of Σ is a substructure of a member of $\mathcal{C}$;

(2) if M and M' are members of $\mathcal{C}$ and M' contains M, then M is existentially complete in M';

(3) if M is a member of Σ, M' is a member of $\mathcal{C}$ and extends M, and M is existentially complete in M', then M is in $\mathcal{C}$.

Proof. The subclass $\mathcal{E}_\Sigma$ of existentially complete structures satisfies (1), (2), and (3) (see Proposition 1.4 and Lemma 1.10). It remains to show that any subclass $\mathcal{D}$ with these properties coincides with $\mathcal{E}_\Sigma$. Since $\mathcal{E}_\Sigma$ contains every class satisfying (1) and (2), one must show only that $\mathcal{E}_\Sigma \subseteq \mathcal{D}$. Let M be a member of $\mathcal{E}_\Sigma$. Since $\mathcal{D}$ satisfies (1), there is a member M' of $\mathcal{D}$ which contains M. Since M is existentially complete in Σ, M is existentially complete in M'. Therefore, by property (3), M is in $\mathcal{D}$. Hence $\mathcal{E}_\Sigma \subseteq \mathcal{D}$, so $\mathcal{E}_\Sigma = \mathcal{D}$.

The assumption in the preceding proposition that Σ is a generalized elementary class is essential. For example, let Σ be the class consisting of (i) the rational numbers Q, (ii) the algebraic closure $\overline{Q}$ of the rationals, and (iii) the function field $Q(x)$ in one variable over the rationals, that is, $\Sigma = \{Q, \overline{Q}, Q(x)\}$. The class Σ is an inductive class but is not a generalized elementary class. $\mathcal{E}_\Sigma$ is the subclass $\{\overline{Q}, Q(x)\}$. $\mathcal{E}_\Sigma$ satisfies the conditions in Proposition 1.4. However, no subclass of Σ satisfies the conditions in Proposition 1.11, as Q is existentially complete in $Q(x)$.

The cardinality of the existentially complete structure constructed in Proposition 1.3 has no a priori bound. A bound does exist when Σ is an inductive, generalized elementary class. Specifically, if Σ is an inductive, generalized elementary class, then each member M of Σ is contained in an existentially complete member M' of Σ such that $\mathrm{card}(M') \le \max\{\mathrm{card}(M), \mathrm{card}(\mathcal{L}(\Sigma)), \aleph_0\}$. This follows from part (1) of Proposition 1.11, the downward Lowenheim-Skolem theorem for generalized elementary classes, and the fact that an elementary substructure of an existentially complete structure is existentially complete also (Lemma 1.10).

Although the class $\Sigma = \mathcal{M}(T_\forall)$ is the natural setting for studying the existentially complete structures for a theory T, it is not the only one. As far as the class $\mathcal{E}_T$ of existentially complete structures is concerned, one may choose $\Sigma = \mathcal{M}(T_{\forall\exists})$ as well as $\Sigma = \mathcal{M}(T_\forall)$.

<u>Lemma 1.12</u>. Let T be a theory and let $\Sigma = \mathcal{M}(T_\forall)$. If M is an existentially complete structure in Σ, then M is a model of $T_{\forall\exists}$.

<u>Proof</u>. Since M is a member of Σ, M is contained in a model M' of T. The lemma now follows from Proposition 1.2.

<u>Proposition 1.13</u>. Let T be a first order theory, and let $\Sigma_1 = \mathcal{M}(T_\forall)$ and $\Sigma_2 = \mathcal{M}(T_{\forall\exists})$. Then $\mathcal{E}_{\Sigma_1} = \mathcal{E}_{\Sigma_2}$.

<u>Proof</u>. Clearly $\Sigma_1 \supseteq \Sigma_2$, and by the preceding lemma $\Sigma_2 \supseteq \mathcal{E}_{\Sigma_1}$. Considered as a subclass of Σ_2, $\mathcal{E}_{\Sigma_1}$ still satisfies conditions (1), (2), and (3) of Proposition 1.11. Therefore, $\mathcal{E}_{\Sigma_1} = \mathcal{E}_{\Sigma_2}$.

The following proposition, a variation of Proposition 1.2, will become the origin of a basic question about the class of existentially complete structures of a theory.

<u>Proposition 1.14</u>. Assume that Σ is a generalized elementary class. Suppose that M and M' are existentially complete members of Σ and that M' extends M. If ϕ is an $\forall\exists\forall$ sentence which is defined in M and is true in M', then ϕ is true in M.

<u>Proof</u>. According to Proposition 1.1, there is an elementary extension M" of M such that

$$\begin{array}{ccc} & & M'' \\ & \preccurlyeq & \\ M' & & \curlyvee \\ & \succcurlyeq & \\ & & M \end{array} \quad .$$

Since ϕ is an $\forall\exists\forall$ sentence, ϕ has the form $\forall v_0 \ldots \forall v_n \exists v_{n+1} \ldots \exists v_m\, \psi(v_0,\ldots, v_n, v_{n+1},\ldots, v_m)$, where ψ is a universal formula with free variables $v_0, \ldots, v_n, v_{n+1}, \ldots, v_m$. Let $r = m - n - 1$. Let $a_0, \ldots, a_n$ be arbitrary but fixed elements of M. Then $\exists v_{n+1} \ldots \exists v_m\, \psi(a_0,\ldots, a_n, v_{n+1},\ldots, v_m)$ is defined in M and is true in M'. Choose elements $b_0, \ldots, b_r$ of M' for which M' satisfies $\psi(a_0,\ldots, a_n, b_0,\ldots, b_r)$.

Since M' is existentially complete in Σ, it is existentially complete in M''. Therefore, M'' satisfies $\psi(a_0,\ldots, a_n, b_0,\ldots, b_r)$ also. So M'' satisfies $\exists v_{n+1} \ldots \exists v_m\, \psi(a_0,\ldots, a_n, v_{n+1},\ldots, v_m)$. Since M'' is an elementary extension of M, M satisfies $\exists v_{n+1} \ldots \exists v_m\, \psi(a_0,\ldots, a_n, v_{n+1},\ldots, v_m)$. Finally, since $a_0, \ldots, a_n$ were arbitrary, M satisfies $\forall v_0 \ldots \forall v_n \exists v_{n+1} \ldots \exists v_m\, \psi(v_0,\ldots, v_n, v_{n+1},\ldots, v_m)$, i.e., M satisfies ϕ.

§ 2 Existential Types and Existentially Universal Structures

Consider again the special case of fields. The algebraic geometry of fields is usually discussed in the context of a universal domain rather than an arbitrary algebraically closed field. A universal domain is an algebraically closed field of infinite transcendence degree over its prime subfield. The reason for the use of universal domains is that every realizable, finitary situation is realized in the universal domain already. For example, if K is a universal domain, k is a finitely generated subfield of K, and L is an extension of k

of finite transcendence degree, then there is a k-isomorphism of L into K.

The goal of this section is the introduction of an analogue for universal domains. The development will parallel that of existentially complete structures.

First, the notion of a finitary situation must be analyzed from the point of view of first order logic. A finitary situation for fields is a finitely generated field. A finitely generated field is completely described by specifying its finite transcendence degree and a finite number of elements which are algebraic over a particular transcendence basis. These algebraic elements are first order definable in terms of the transcendence basis. One merely specifies their irreducible polynomials. However, the elements of the transcendence basis are not first order definable themselves. To specify that an element is transcendental requires an infinite number of sentences, for each sentence can assert only that the element is not a zero of each of a finite number of polynomials.

The concept of an existential type is introduced to handle transcendentals. Let Σ be a class of similar structures, and let $\mathcal{L}(\Sigma)$ be a language appropriate for Σ. An <u>existential</u> <u>n-type</u> (for Σ) is a set Δ of existential formulas of $\mathcal{L}(\Sigma)$ with free variables among $v_0, \ldots, v_{n-1}$. The notation $\Delta(v_0,\ldots, v_{n-1})$ will be used to indicate the free variables. Although this concept and all others introduced in this section are relative to the class Σ and the language $\mathcal{L}(\Sigma)$, this dependency will not be stated explicitly in all instances.

If $\Delta(v_0,\ldots, v_{n-1})$ is an existential n-type and a_i for some $i < n$ is a constant symbol (not necessarily in $\mathcal{L}(\Sigma)$), then $\Delta(v_0,\ldots, v_{i-1}, a_i, v_{i+1},\ldots, v_{n-1})$ will denote the set of sentences obtained by substituting the constant symbol a_i for each free occurrence of the variable v_i in each formula in $\Delta(v_0,\ldots, v_{n-1})$.

A structure M in Σ is said to realize an existential n-type $\Delta(v_0, \ldots, v_{n-1})$ if there are elements $a_0, \ldots, a_{n-1}$ in M such that M satisfies $\phi(a_0, \ldots, a_{n-1})$ for each formula $\phi(v_0, \ldots, v_{n-1})$ in Δ. Alternatively, one says that $a_0, \ldots, a_{n-1}$ realize $\Delta(v_0, \ldots, v_{n-1})$ in M or that M satisfies $\Delta(a_0, \ldots, a_{n-1})$.

If M is a member of Σ and $a_0, \ldots, a_{n-1}$ are elements of M, then the existential type of $a_0, \ldots, a_{n-1}$ in M is the set of all existential formulas $\phi(v_0, \ldots, v_{n-1})$ in the language $\mathcal{L}(\Sigma)$ with free variables among $v_0, \ldots, v_{n-1}$ such that M satisfies $\phi(a_0, \ldots, a_{n-1})$. For example, let t be a transcendental element over the rational numbers Q. Then the existential 1-type of t in $Q(t)$ includes a formula $\phi(v_0)$ for each polynomial $p(x)$ with coefficients from Q such that $\phi(t)$ asserts that t is not a zero of $p(x)$. If M and M' are members of Σ, M is contained in M', and $a_0, \ldots, a_{n-1}$ are elements of M, then the existential type of $a_0, \ldots, a_{n-1}$ in M is contained in the existential type of $a_0, \ldots, a_{n-1}$ in M'. This is because true existential sentences are persistent under extension.

An existential n-type Δ is said to be maximal if (i) there is a structure M in Σ which realizes Δ, and (ii) whenever $\phi(v_0, \ldots, v_{n-1})$ is an existential formula in the language $\mathcal{L}(\Sigma)$ and $\Delta' = \Delta \cup \{\phi(v_0, \ldots, v_{n-1})\} \neq \Delta$, then no member M' of Σ realizes Δ'.

Let $\Delta(v_0, \ldots, v_{n-1}, v_n, \ldots, v_{n+r})$ be an existential $(n + r + 1)$-type. Let M be a member of Σ and let $b_0, \ldots, b_r$ be elements of M. Then $\Delta(v_0, \ldots, v_{n-1}, b_0, \ldots, b_r)$ is said to be an existential n-type defined in M.

The phrase "... existential type ..." may be used in place of the phrase "... existential n-type ..." whenever the number n is clear from context, as in $\Delta(v_0, \ldots, v_{n-1})$ for example, or whenever the only significant attribute of n is its being a positive integer.

A structure M in Σ is said to be <u>existentially universal</u> (in Σ) if each existential type which is defined in M and is realized in some extension of M in Σ is realized in M itself. For example, any universal domain is existentially universal in the class of fields. The proof is as follows. Let K be a universal domain, and let $\Delta(v_0,\ldots, v_{n-1}, b_0,\ldots, b_r)$ be an existential type defined in K, where $b_0,\ldots, b_r$ are elements of K. Suppose that L is a field extension of K and that $a'_0, \ldots, a'_{n-1}$ are elements of L such that L satisfies $\Delta(a'_0,\ldots, a'_{n-1}, b_0,\ldots, b_r)$. Let k be the prime subfield of K. Let $\bar{L}$ be the algebraic closure of L and let F' be the algebraic closure of the field $k(a'_0,\ldots, a'_{n-1}, b_0,\ldots, b_r)$ in $\bar{L}$. Since K is a universal domain, there are elements $a_0, \ldots, a_{n-1}$ in K for which there is a $k(b_0,\ldots, b_r)$-isomorphism of F' onto the algebraic closure F of $k(a_0,\ldots, a_{n-1}, b_0,\ldots, b_r)$ in K. Hence K satisfies $\Delta(a_0,\ldots, a_{n-1}, b_0,\ldots, b_r)$. Conversely, an existentially universal structure in the class of fields is a universal domain.

The subclass of existentially universal members of Σ will be denoted by $\mathcal{A}_\Sigma$. Note that $\mathcal{A}_\Sigma$ is contained in $\mathcal{E}_\Sigma$, for suppose M is existentially universal. Suppose further that ϕ is an existential sentence which is defined in M and is true in some extension M' of M in Σ. Then ϕ has the form $\exists v_0 \ldots \exists v_{n-1}\, \psi(v_0,\ldots, v_{n-1})$, where ψ is a quantifier-free formula defined in M. Then $\Delta = \{\psi(v_0,\ldots, v_{n-1})\}$ is an existential type defined in M and realized in M', so it is realized in M also.

Existentially universal structures exist whenever Σ is an inductive class.

<u>Proposition 1.15</u>. Assume Σ is an inductive class. Then each member of Σ is a substructure of an existentially universal member of Σ.

Proof. The proof is almost identical to that of Proposition 1.3 and so will be omitted. The principal change is that existential types replace existential sentences in the argument. One also needs the observation that for any structure M in Σ, the number of existential types defined in M is less than or equal to
$\mathrm{card}(M) \cdot 2^{\aleph_0} \cdot \mathrm{card}(\mathcal{L}(\Sigma))$.

The class of existentially universal structures has a characterization similar to that of the class of existentially complete structures.

Proposition 1.16. Assume Σ is an inductive class. Then $\mathcal{A}_\Sigma$ is the unique subclass $\mathcal{C}$ of Σ satisfying

(1) every member of Σ is a substructure of a member of $\mathcal{C}$;

(2) if M and M' are members of $\mathcal{C}$ and M' extends M, then any existential type defined in M and realized in M' is realized in M itself;

(3) $\mathcal{C}$ contains every other subclass of Σ which satisfies conditions (1) and (2).

Proof. The class $\mathcal{A}_\Sigma$ satisfies (1) and (2), so it suffices to show that each subclass satisfying (1) and (2) is contained in $\mathcal{A}_\Sigma$. Suppose $\mathcal{D}$ is a subclass of Σ and satisfies (1) and (2). Let M be a member of $\mathcal{D}$. Suppose that Δ is an existential type which is defined in M and is realized in some extension M' of M in Σ. Since $\mathcal{D}$ satisfies (1), there is an M'' in $\mathcal{D}$ containing M'. Since existential types are persistent under extension, M'' realizes Δ. Finally, since $\mathcal{D}$ satisfies (2), M must realize Δ. Therefore, M is existentially universal.

Proposition 1.17. Assume Σ is an inductive class. Then the class $\mathcal{A}_\Sigma$ is inductive.

Proof. Let $\{M_\alpha : \alpha < \lambda\}$ be an ascending chain of members of $\mathcal{A}_\Sigma$, and let $M = \bigcup_{\alpha<\lambda} M_\alpha$. Let Δ be an existential type defined in M. Since Δ mentions only finitely many elements of M, Δ is defined in M_α for some $\alpha < \lambda$. So, if Δ is realized in some extension of M in Σ, then Δ is realized in M_α and consequently is realized in M. Thus, M is existentially universal.

Further results depend upon the assumption that Σ is a generalized elementary class. Consequences of this assumption are a simpler criterion for a structure to be existentially universal, an alternative form of condition (3) in Proposition 1.16, an amalgamation property for existential types and a characterization of the existential types of elements in existentially complete structures, and the fact that all elementary properties in an existentially universal structure are determined entirely by existential properties. Although most of these results are valid for arbitrary, generalized elementary classes, nevertheless one should think in terms of inductive, generalized elementary classes. Otherwise, the results may be true vacuously.

The first step towards a simpler criterion for a structure to be existentially universal is the elimination of the reference to existential types realized in extensions. Assume that Σ is a generalized elementary class, and let T_Σ be a set of axioms for Σ. Suppose that M is a member of Σ and that $\Delta(v_0,\ldots, v_{n-1}, b_0,\ldots, b_r)$ is an existential type defined in M. Let $c_0, \ldots, c_{n-1}$ be new constant symbols which have not been assigned as names of elements of M. The existential type $\Delta(v_0,\ldots, v_{n-1}, b_0,\ldots, b_r)$ is said to be <u>finitely consistent with M</u> if each finite subset of $\Delta(c_0,\ldots, c_{n-1}, b_0,\ldots, b_r)$ is consistent with $T_\Sigma \cup \mathrm{Diag}(M)$. By the completeness theorem, $\Delta(v_0,\ldots, v_{n-1}, b_0,\ldots, b_r)$ is finitely consistent with M if and only if every finite subset of $\Delta(v_0, \ldots, v_{n-1}, b_0, \ldots, b_r)$ is realized in some extension of M in Σ. By the compactness theorem, an existential type which is finitely consistent with M is realized in some extension of M in Σ. This proves the following lemma.

Lemma 1.18. Assume Σ is a generalized elementary class. An existential type defined in a member M of Σ is realized in some extension of M in Σ if and only if it is finitely consistent with M.

Lemma 1.19. Assume Σ is a generalized elementary class. A member M of Σ is existentially universal if and only if every existential type defined in M and finitely consistent with M is realized in M.

Proof. Immediate from the definitions and the preceding lemma.

The next step is to restrict consideration to existential 1-types only.

An existential type $\Delta(v_0, \ldots, v_{n-1}, b_0, \ldots, b_r)$ defined in a structure M is said to be *finitely satisfiable in M* if each finite subset of $\Delta(v_0, \ldots, v_{n-1}, b_0, \ldots, b_r)$ is realized in M.

Proposition 1.20. Assume Σ is a generalized elementary class. A member M of Σ is existentially universal if and only if every existential 1-type defined in M and realized in some extension of M in Σ is realized in M itself.

Proof. Necessity is obvious.

Assume that M satisfies the second condition. Let $\Delta(v_0, \ldots, v_{n-1}, b_0, \ldots, b_r)$ be an existential type defined in M and realized in some extension M' of M in Σ. Without loss of generality, we may assume that the conjunction of any finite number of formulas from Δ is also in Δ. If $\phi(v_0, \ldots, v_{n-1})$ is a formula in Δ, let ϕ' be the formula $\exists v_1 \ldots \exists v_{n-1}\, \phi(v_0, \ldots, v_{n-1})$.

Let $\Delta'(v_0)$ be the set $\{\phi' : \phi \in \Delta\}$. Then $\Delta'(v_0)$ is an existential 1-type defined in M and realized in M'. By assumption, $\Delta'(v_0)$ is realized in M by some element a_0 of M.
Let $\Delta_1 = (a_0, v_1, \ldots, v_{n-1}, b_0, \ldots, b_r)$. This is an existential type defined in M. Now Δ_1 is finitely satisfiable in M, because $\Delta'(a_0)$ is satisfied in M; so, Δ_1 is finitely consistent with M. Consequently, Δ is realized in some extension M' of M in Σ.

Continuing in this manner, one obtains elements $a_0, a_1, \ldots, a_{n-1}$ of M which realize Δ. Hence, M is existentially universal.

The next lemma follows immediately from definitions and Proposition 1.6.

Lemma 1.21. Assume Σ is a generalized elementary class. Suppose M is an existentially complete member of Σ. Then

(1) an existential type defined in M is finitely consistent with M if and only if it is finitely satisfiable in M; and

(2) an existential type defined in M is realized in some extension of M if and only if it is finitely satisfiable in M.

Proposition 1.22. Assume Σ is a generalized elementary class. Suppose M is an existentially complete member of Σ. Then M is existentially universal if and only if each existential 1-type defined in M and finitely satisfiable in M is realized in M.

Proof. According to Proposition 1.20, it suffices to consider only existential 1-types. Assume M is existentially universal. If Δ is an existential 1-type defined in M and finitely satisfiable in M, then Δ is finitely consistent with M and so is realized in some extension of M in Σ. Therefore, Δ is realized in M.

Conversely, assume the second condtion. Let Δ be an existential 1-type defined in M and realized in some extension of M in Σ.

Since M is existentially complete, Δ is finitely satisfiable in M. Hence, Δ is realized in M.

The next objective is the replacement of the maximality condition in Proposition 1.16 by a closure condition. The amalgamation property for existentially complete structures (Corollary 1.9) is necessary for this replacement. One should recall the example after Proposition 1.11 in which $\Sigma = \{Q, \overline{Q}, Q(x)\}$. In this example the structures $\overline{Q}$ and $Q(x)$ cannot be amalgamated over Q in Σ. Consequently, although Q is existentially complete in the existentially complete structure $Q(x)$ of Σ, Q is not itself existentially complete in Σ. The same difficulty can arise for existentially universal structures.

<u>Lemma 1.23</u>. Assume Σ is a generalized elementary class. Suppose M and M' are members of Σ such that M' is existentially universal, M' extends M, and any existential type which is defined in M and is realized in M' is realized in M. Then M is existentially universal also.

<u>Proof</u>. Let $\Delta(v_0, \ldots, v_{n-1}, b_0, \ldots, b_r)$ be an existential type defined in M and realized in some extension M'' of M in Σ. The hypotheses imply that M is existentially complete. Therefore, there is a structure M''' in Σ such that

$$\begin{array}{ccccc} & & M''' & & \\ & \subseteq & & \supseteq & \\ M' & & & & M'' \\ & \supseteq & & \subseteq & \\ & & M & & \end{array} .$$

Since existential types are persistent under extension, Δ is realized in M''' . Consequently, Δ is realized in M' and so in M itself. Therefore, M is existentially universal.

Theorem 1.24. Assume Σ is an inductive, generalized elementary class. Then $\mathcal{A}_\Sigma$ is the unique subclass $\mathcal{C}$ of Σ satisfying

(1) each member of Σ is contained in a member of $\mathcal{C}$;

(2) if M and M' are members of $\mathcal{C}$ and M' contains M, then any existential type defined in M and realized in M' is realized in M;

(3) if M is a member of Σ, M' is a member of $\mathcal{C}$ and contains M, and any existential type defined in M and realized in M' is realized in M, then M is in $\mathcal{C}$ also.

Proof. The proof is almost identical to that of Proposition 1.11 and will be omitted.

In order to characterize the existential types of elements in existentially complete and existentially universal structures, one must know that structures can be "glued together" along sequences of elements which realize mutually consistent existential types.

Lemma 1.25. Assume Σ is a generalized elementary class. Let M and M' be members of Σ, and let $a_0, \ldots, a_{n-1}$ and $b_0, \ldots, b_{n-1}$ be elements of M and M' respectively, The following are equivalent:

(i) there is a structure M'' in Σ such that M'' contains M and the existential type of $b_0, \ldots, b_{n-1}$ in M' is contained in the existential type of $a_0, \ldots, a_{n-1}$ in M'';

(ii) there is a structure M'' in Σ which contains both M and M' and in which $a_0, \ldots, a_{n-1}$ and $b_0, \ldots, b_{n-1}$ are identified, i.e., $a_i = b_i$ for $i = 0, \ldots, n-1$ in M''. Diagramatically,

$$\begin{array}{ccccc} & & M'' & & \\ & \subseteq & & \supseteq & \\ M & & & & M' \\ \uparrow & & & & \uparrow \\ (a_0,\ldots, a_{n-1}) & & \longleftrightarrow & & (b_0,\ldots, b_{n-1}) \end{array} .$$

Proof. The second condition clearly implies the first.

Conversely, assume that the first condition is satisfied. Let $\Delta(v_0,\ldots, v_{n-1})$ be the existential type of $a_0, \ldots, a_{n-1}$ in M''. Let T_Σ be a set of axioms for Σ, and let Diag(M) and Diag(M') be the diagrams of M and M', respectively, where the only constant symbols occurring in both Diag(M) and Diag(M') are the constant symbols of $\mathcal{L}(\Sigma)$. Let MDiag(M') be the set of sentences obtained by replacing each occurrence of b_i in Diag(M') be an occurrence of a_i for $i = 0, \ldots, n-1$. The second condition is satisfied if and only if $T_\Sigma \cup \text{Diag}(M) \cup \text{MDiag}(M')$ is a consistent set of sentences. Let $\phi(a_0,\ldots, a_{n-1}, c_0,\ldots, c_r)$ be a conjunction of a finite number of sentences from Diag(M), and let $\psi(a_0,\ldots, a_{n-1}, d_0,\ldots, d_s)$ be a conjunction of a finite number of sentences from MDiag(M'), where $c_0, \ldots, c_r$ and $d_0, \ldots, d_s$ are the constants occurring in ϕ and ψ other than $a_0, \ldots, a_{n-1}$ and the constants of $\mathcal{L}(\Sigma)$. Then the formula $\exists v_n \ldots \exists v_{n+r}\, \phi(v_0,\ldots, v_{n-1}, v_n,\ldots, v_{n+r})$ is in the existential type of $a_0, \ldots, a_{n-1}$ in M. This formula is also in the existential type $\Delta(v_0,\ldots, v_{n-1})$ of $a_0, \ldots, a_{n-1}$ in M'', because true existential sentences are persistent under extension. The formula $\exists v_n \ldots \exists v_{n+s}\, \psi(v_0,\ldots, v_{n-1}, v_n,\ldots, v_{n+s})$ is in the existential type of $b_0, \ldots, b_{n-1}$ in M', so by assumption it is in the existential type $\Delta(v_0,\ldots, v_{n-1})$. Since Δ is closed under conjunction of formulas from Δ, the formula

$$\exists v_n \ldots \exists v_{n+r}\, \phi(v_0,\ldots, v_{n-1}, v_n,\ldots, v_{n+r})$$
$$\wedge \exists v_n \ldots \exists v_{n+s}\, \psi(v_0,\ldots, v_{n-1}, v_n,\ldots, v_{n+s})$$

is in $\Delta(v_0,\ldots, v_{n-1})$. Therefore, $T_\Sigma \cup \text{Diag}(M) \cup \text{MDiag}(M')$ is consistent.

Corollary 1.26. Assume Σ is a generalized elementary class. If M and M' are members of Σ, and $a_0, \ldots, a_{n-1}$ and $b_0, \ldots, b_{n-1}$ are elements of M and M', respectively, which have the same existential type in M and M', respectively, then there is a structure M'' in Σ such that the following diagram commutes:

$$\begin{array}{ccccc} & & M'' & & \\ & \nearrow & & \nwarrow & \\ M & & & & M' \\ \uparrow & & & & \uparrow \\ (a_0, \ldots, a_{n-1}) & & \longleftrightarrow & & (b_0, \ldots, b_{n-1}) \end{array} .$$

Corollary 1.27. Assume Σ is a generalized elementary class. Let $\Delta(v_0, \ldots, v_{n-1})$ be a maximal existential type for Σ. If M and M' are members of Σ, and $a_0, \ldots, a_{n-1}$ and $b_0, \ldots, b_{n-1}$ are elements of M and M', respectively, which realize Δ in M and M', respectively, then there is a structure M'' in Σ such that the following diagram commutes:

$$\begin{array}{ccccc} & & M'' & & \\ & \nearrow & & \nwarrow & \\ M & & & & M' \\ \uparrow & & & & \uparrow \\ (a_0, \ldots, a_{n-1}) & & \longleftrightarrow & & (b_0, \ldots, b_{n-1}) \end{array} .$$

Amalgamation over existential types yields the following characterization of existentially complete structures and the existential types of elements in existentially complete structures.

Proposition 1.28. Assume Σ is a generalized elementary class. A structure in Σ is existentially complete if and only if the existential type of each finite sequence of elements in that structure is a maximal existential type for Σ.

Proof. First, assume M is an existentially complete structure in Σ. Let $a_0, \ldots, a_{n-1}$ be elements of M, and let $\Delta(v_0, \ldots, v_{n-1})$ be the existential type of $a_0, \ldots, a_{n-1}$ in M. Let $\phi(v_0, \ldots, v_{n-1})$

be an existential formula in $\mathcal{L}(\Sigma)$, and suppose there is a member M' of Σ with elements $b_0, \ldots, b_{n-1}$ which realize the existential type $\Delta'(v_0,\ldots, v_{n-1}) = \Delta(v_0,\ldots, v_{n-1}) \cup \{\phi(v_0,\ldots, v_{n-1})\}$. Then the first condition in Lemma 1.25 is satisfied, so there is an M'' in Σ such that the following diagram commutes:

$$\begin{array}{ccc} & M'' & \\ \nearrow & & \nwarrow \\ M & & M' \\ \uparrow & & \uparrow \\ (a_0,\ldots, a_{n-1}) & \longleftrightarrow & (b_0,\ldots, b_{n-1}) \end{array}.$$

M'' satisfies $\phi(a_0,\ldots, a_{n-1})$, since M' satisfies $\phi(b_0,\ldots, b_{n-1})$ and $a_i = b_i$ for $i = 0, \ldots, n-1$. Since M is existentially complete, M satisfies $\phi(a_0,\ldots, a_{n-1})$, so $\phi(v_0,\ldots, v_{n-1})$ is in Δ. Hence, Δ is a maximal existential type.

Conversely, assume the existential type of each finite sequence of elements in a member M of Σ is a maximal existential type. Let $\phi(v_0,\ldots, v_{n-1})$ be an existential formula in $\mathcal{L}(\Sigma)$. Suppose $\phi(a_0,\ldots, a_{n-1})$ is defined in M and true in some extension M' of M in Σ. Then the existential type of $a_0, \ldots, a_{n-1}$ in M' includes both $\phi(v_0,\ldots, v_{n-1})$ and all of the formulas in the existential type $\Delta(v_0,\ldots, v_{n-1})$ of $a_0, \ldots, a_{n-1}$ in M. By assumption, Δ is a maximal existential type, so Δ includes $\phi(v_0,\ldots, v_{n-1})$. Thus, M satisfies $\phi(a_0,\ldots, a_{n-1})$.

Corollary 1.29. Assume Σ is an inductive, generalized elementary class. Then an existential type is maximal if and only if it is the existential type of a sequence of elements in some existentially complete structure.

An important property of existentially universal structures in a generalized elementary class is that everything of an elementary nature is determined entirely by the maximal existential types which are realized. This statement is explicated in the remainder of this section.

A complete n-type for Σ is a set $\Theta(v_0,\ldots, v_{n-1})$ of formulas in the language $\mathcal{L}(\Sigma)$ with free variables among $v_0, \ldots, v_{n-1}$ such that for each formula $\phi(v_0,\ldots, v_{n-1})$ in $\mathcal{L}(\Sigma)$ either ϕ or $\sim\phi$ is in Θ. The complete type $\Theta(v_0,\ldots, v_{n-1})$ of the elements $a_0, \ldots, a_{n-1}$ in a structure $\underline{M}$ in Σ is the set of all formulas $\phi(v_0,\ldots, v_{n-1})$ in $\mathcal{L}(\Sigma)$ such that M satisfies $\phi(a_0,\ldots, a_{n-1})$. The definitions of a complete type defined in a structure $\underline{M}$ and of $\underline{M}$ realize a complete type $\underline{\Theta}$ are analogous to the corresponding definitions for existential types.

Proposition 1.30. Assume Σ is a generalized elementary class. Suppose that M and M' are existentially universal members of Σ and that $a_0, \ldots, a_{n-1}$ and $b_0, \ldots, b_{n-1}$ are elements of M and M', respectively, which realize the same maximal existential type in M and M', respectively. Then $a_0, \ldots, a_{n-1}$ and $b_0, \ldots, b_{n-1}$ realize the same complete type in M and M', respectively.

Proof. By a standard inductive argument (see Karp (51)) and by the symmetry between M and $a_0, \ldots, a_{n-1}$ and M' and $b_0, \ldots, b_{n-1}$, it suffices to show that if $c_0, \ldots, c_r$ are elements of M, then there are elements $d_0, \ldots, d_r$ of M' such that the substructure of M generated by $a_0, \ldots, a_{n-1}, c_0, \ldots, c_r$ is isomorphic by the natural isomorphism to the substructure of M' generated by $b_0, \ldots, b_{n-1}, d_0, \ldots, d_r$. Since $a_0, \ldots, a_{n-1}$ and $b_0, \ldots, b_{n-1}$ realize the same maximal existential type in M and M', respectively, there is an M'' in Σ such that the following diagram commutes:

$$\begin{array}{ccccc} & & M'' & & \\ & \swarrow\!\!\!\!\!\subseteq & & \supseteq\!\!\!\!\!\searrow & \\ M & & & & M' \\ \uparrow & & & & \uparrow \\ (a_0,\ldots, a_{n-1}) & & \longleftrightarrow & & (b_0,\ldots, b_{n-1}) \end{array} .$$

Suppose $c_0, \ldots, c_r$ are elements of M. Let $\Delta(v_0,\ldots, v_{n-1}, v_n,\ldots, v_{n+r})$ be the existential type of $a_0, \ldots, a_{n-1}, c_0, \ldots, c_r$ in M. Then $\Delta(b_0,\ldots, b_{n-1}, v_n,\ldots, v_{n+r})$ is an existential type defined in M' and realized in M'' by $b_0, \ldots, b_{n-1}, c_0, \ldots, c_r$. Since M' is existentially universal, there are elements $d_0, \ldots, d_r$ in M' such that $b_0, \ldots, b_{n-1}, d_0, \ldots, d_r$ realize Δ in M'. Since the substructure generated by a set of elements is completely described in their existential type, the substructures generated by $a_0, \ldots, a_{n-1}, c_0, \ldots, c_r$ in M and by $b_0, \ldots, b_{n-1}, d_0, \ldots, d_r$ in M' are isomorphic.

Proposition 1.31. Assume Σ is a generalized elementary class. Suppose M and M' are existentially universal structures in Σ such that M' contains M. Then

(1) each finite sequence of elements of M realizes the same complete type in both M and M';

(2) each complete type defined in M and realized in M' is realized in M itself;

(3) $M \prec_{\infty,\omega} M'$.

Proof. (1) Let $a_0, \ldots, a_{n-1}$ be elements of M. Since M is existentially complete, these elements realize the same maximal existential type in both M and M'. According to the preceding proposition, $a_0, \ldots, a_{n-1}$ realize the same complete type in both M and M'.

(2) Let $\Theta(v_0,\ldots, v_{n-1}, v_n,\ldots, v_{n+r})$ be a complete type for Σ, and suppose $b_0, \ldots, b_r$ are elements of M such that the

type $\Theta(v_0,\ldots, v_{n-1}, b_0,\ldots, b_r)$ is realized in M'. Let $\Delta(v_0,\ldots, v_{n-1}, v_n,\ldots, v_{n+r})$ be the set of existential formulas in Θ. Then $\Delta(v_0,\ldots, v_{n-1}, b_0,\ldots, b_r)$ is an existential type defined in M and realized in M', so there are elements $a_0, \ldots, a_{n-1}$ of M which realize $\Delta(v_0,\ldots, v_{n-1}, b_0,\ldots, b_r)$. According to the preceding proposition, $a_0, \ldots, a_{n-1}, b_0,\ldots, b_r$ realize Θ in M.

(3) This part follows immediately from parts (1) and (2).

Proposition 1.32. Assume Σ is a generalized elementary class. If two existentially universal structures M and M' in Σ satisfy the same existential sentences in the language of Σ, then

(1) each existential type realized in one is realized in the other also;

(2) each complete type realized in one is realized in the other also;

(3) $M \equiv_{\infty,\omega} M'$.

Proof. (1) Since M and M' satisfy the same existential sentences, any existential type realized in one is finitely satisfiable in the other and hence, is realized in the other (Lemma 1.19 and Lemma 1.21.).

(2) This part follows from part (1) and Proposition 1.30.

(3) This part follows from part (2).

Proposition 1.33. Assume Σ is a generalized elementary class. If M and M' are existentially universal structures in Σ which satisfy the same existential sentences of $\mathcal{L}(\Sigma)$, and M is countable, then M can be embedded in M'. If in addition M' is countable also, then M and M' are isomorphic.

Proof. This follows from part (3) of the preceding proposition. Alternatively, one can construct the embedding (or isomorphism) using Lemma 1.19 and Lemma 1.21 (and a back and forth construction).

CHAPTER 2

MODEL-COMPLETIONS AND MODEL-COMPANIONS

The first metamathematical analogue of algebraically closed fields was formal rather than material as in the previous chapter. The formal analogue is based upon the fact that algebraically closed fields are more than just existentially complete in one another. If K and L are algebraically closed fields and L is an extension of K, then a sentence defined in K is true in K if and only if it is true in L. Moreover, a sentence defined in a field k is true in one algebraically closed extension of k if and only if it is true in all algebraically closed extensions of k. These properties were the motivation for A. Robinson's concepts of model-completeness and model-completion. The purpose of this chapter is to review these formal analogues and relate them to the first chapter.

The notions of model-consistency and model-completeness can be defined either syntactically or semantically. We shall state the semantic definitions and then mention the equivalent syntactic requirements. Let $\mathcal{L}$ be a first order language, and let T and T^* be two theories in $\mathcal{L}$. The theory T^* is said to be <u>model-consistent relative to T</u> if each model of T is contained in a model of T^*. Equivalent syntactic formulations are (i) for each model M of T the set of sentences $T^* \cup \mathrm{Diag}(M)$ is consistent or (ii) every universal sentence deducible from T^* is also deducible from T, that is, $T_\forall$ contains $(T^*)_\forall$. The theories T and T^* are said to be <u>mutually model-consistent</u> if each is model-consistent relative to the other, that is, each model of one theory is contained in a model of the other theory. Equivalently, T and T^* are mutually model-consistent if and only if $T_\forall = (T^*)_\forall$.

The theory T^* is said to be <u>model-complete relative to T</u> if whenever M is a model of T, M' and M'' are models of T^* and contain M as a substructure, and ϕ is a sentence defined in M, then M' satisfies ϕ if and only if M'' satisfies ϕ. Equivalently, T^* is model-complete relative to T if and only if for each model M of T the theory $T^* \cup \mathrm{Diag}(M)$ is a complete theory in the language of M. If T^* is model-complete relative to itself, then T^* is said to be <u>model-complete</u>. In other words, T^* is model-complete if and only if whenever one model of T^* is an extension of a second model of T^*, then the first model is an elementary extension of the second model. Finally, if T^* contains T, is model-consistent relative to T, and is model-complete relative to T, then T^* is called a <u>model-completion of T</u>. Note that if T^* is a model-completion, then T^* is model-complete itself.

In the case of fields, let T be a set of axioms for the class of fields, and let T^* be a set of axioms for the class of algebraically closed fields. The theory T^* is model-consistent relative to the theory T, for every field is contained in an algebraically closed field. Moreover, T^* is model-complete relative to T. Since T^* contains T, T^* is a model-completion of T (in fact, T^* is <u>the</u> model-completion of T, as will be noted below). Also, since each model of T^* is a field, T^* is model-complete relative to itself, so T^* is a model-complete theory.

There are two important theorems on model-completions and model-completeness. The first concerns the uniqueness of a model-completion, and the second is Robinson's model-completeness test.

<u>Theorem 2.1</u>. If each of two theories is a model-completion for the same, third theory, then the two model-completions are logically equivalent, and each model of one is also a model of the other.

Proof. See Robinson (79), page 109, or the proofs of Theorem 3.8 and Theorem 3.9, or Theorem 2.4.

Accordingly, one refers to the model-completion of a theory, if one exists.

The following theorem is essentially Robinson's model-completeness test.

Theorem 2.2. A theory is model-complete if and only if whenever one model of the theory is a substructure of a second model of the theory, then the first model is existentially complete in the second model.

Proof. Let T be a theory. Assume T is model-complete, and let M_1 and M_2 be models of T such that M_1 is a substructure of M_2. Then any sentence defined in M_1 is true in M_1 if and only if it is true in M_2. As this includes existential sentences, M_1 is existentially complete in M_2.

Conversely, assume that whenever one model of T is a substructure of a second model of T, then the first model is existentially complete in the second model. Let M_1 and M_2 be models of T such that M_1 is a substructure of M_2. Since M_1 is existentially complete in M_2, there is an elementary extension M_3 of M_1 (see Proposition 1.1) for which the following diagram commutes:

$$\begin{array}{ccccc} M_1 & & \prec & & M_3 \\ & \subseteq & & \subseteq & \\ & & M_2 & & \end{array} .$$

The structure M_3 is a model of T, so M_2 is existentially complete in M_3. Therefore, there is an elementary extension M_4 of M_2 such that the following diagram commutes:

$$\begin{array}{ccccccc} M_1 & & \prec & & M_3 & & \\ & \subseteq & & \subseteq & & \subseteq & \\ & & M_2 & & \prec & & M_4 \end{array} .$$

Continuing in this manner, one obtains an infinite chain $\{M_n : n = 1, 2, \ldots\}$ of models of T such that the following diagram commutes:

$$\begin{array}{ccccccccccccc} M_1 & & \prec & & M_3 & & \prec & & M_5 & & \prec & & \\ & \subseteq & & \subseteq & & \subseteq & & \subseteq & & \subseteq & & \subseteq & \cdots \\ & & M_2 & & \prec & & M_4 & & \prec & & M_6 & & \end{array} .$$

Let $M_\omega = \bigcup_{n<\omega} M_{2n+2} = \bigcup_{n<\omega} M_{2n+1}$. Then the following diagram commutes:

$$\begin{array}{ccccc} M_1 & & \prec & & M_\omega \\ & \subseteq & & \succ & \\ & & M_2 & & \end{array} .$$

Therefore, M_1 is an elementary substructure of M_2.

The relationship between model-completeness and existential completeness will be discussed again shortly. The following proposition is included for use later.

<u>Proposition 2.3</u>. If a theory is model-complete, then it can be axiomatized by $\forall\exists$ sentences.

<u>Proof</u>. Whenever a model of a model-complete theory is a substructure of another model of the theory, the first model is an elementary substructure of the second model. Tarski's elementary chain principle implies that this theory is an inductive theory. Therefore, it can be axiomatized by $\forall\exists$ sentences.

There are other important examples of model-completions besides algebraically closed fields. The theory of real closed ordered fields is the model-completion of the theory of ordered fields. The theory of differentially closed fields is the model-completion of the theory of differential fields. The theory of Hensel fields with a discrete valuation is the model-completion of the theory of discrete, nonarchimedean valued fields. For other examples, please consult the chart at the end of Part One.

However, at least one important theory, the theory of formally real fields, does not have a model-completion. The natural candidate for a model-completion for the theory of formally real fields would be the theory of real closed fields. The theory of real closed fields is model-consistent relative to the theory of formally real fields and is model-complete also. But the theory of real closed fields is not model-complete relative to the theory of formally real fields. The reason is the multiplicity of possible orderings for formally real fields. For example, in the field $Q(\sqrt{2})$, either $\sqrt{2}$ or $-\sqrt{2}$ may be taken to be positive. Consequently, in one real closure of $Q(\sqrt{2})$ the element $\sqrt{2}$ has a square root, while in another real closure it does not have a square root. Thus, there are two real closed fields which are extensions of $Q(\sqrt{2})$ and which are not elementarily equivalent in the language of $Q(\sqrt{2})$. Therefore, the theory of real closed fields is not model-complete relative to the theory of formally real fields.

This example motivated the concept of a model-companion, introduced by Eli Bers in 1969. Let T and T^* be theories in a common language. The theory T^* is called a _model-companion for T_ if (i) T and T^* are mutually model-consistent and (ii) T^* is model-complete. Every model-completion is also a model-companion but not _vice versa_, as the example of formally real fields shows. Model-companions and the class of existentially complete structures are related as follows.

Theorem 2.4. Let T be a first order theory.

(1) If T^* is a model-companion for T, then the class of models of T^* is precisely the class $\mathcal{E}_T$ of existentially complete structures for T, and $\mathfrak{Th}(\mathcal{E}_T)$ is the deductive closure of T^*.

(2) If each of two theories is a model-companion for T, then

they are logically equivalent.

(3) If $\mathcal{E}_T$ is a generalized elementary class, then $\mathcal{TA}(\mathcal{E}_T)$ is a model-companion for T.

Proof. The proof of part (1) consists of verifying that the class $\mathcal{M}(T^*)$ satisfies the conditions of Proposition 1.11. Since T and T* are mutually model-consistent, they have the same universal consequences, so $\mathcal{M}(T^*)$ is a subclass of $\Sigma_T = \mathcal{M}(T_\forall)$. First, every structure in Σ_T is contained in a model of T*, since T* is model-consistent relative to T. Secondly, because T* is model-complete, it is true that whenever M and M' are models of T* and M is a substructure of M', then M is existentially complete in M'. It remains to show that if M is in Σ_T, M' is a model of T* which extends M, and M is existentially complete in M', then M is a model of T* also. According to Proposition 2.3, T* can be axiomatized by $\forall\exists$ sentences. Now since M is existentially complete in M', M satisfies all the $\forall\exists$ sentences in the language of T* that M' does; so, M is a model of T*. Thus, $\mathcal{M}(T^*) = \mathcal{E}_T$, and $\mathcal{TA}(\mathcal{E}_T)$ is the deductive closure of T*.

(2) If T_1 and T_2 are both model-companions of T, then according to part (1) $\mathcal{M}(T_1) = \mathcal{E}_T = \mathcal{M}(T_2)$, so T_1 and T_2 are logically equivalent.

(3) Suppose that $\mathcal{E}_T$ is a generalized elementary class, that is, if $T^* = \mathcal{TA}(\mathcal{E}_T)$, then $\mathcal{E}_T = \mathcal{M}(T^*)$. Since each structure in $\mathcal{E}_T$ is a substructure of a model of T, and since each model of T is contained in a member of $\mathcal{E}_T$, the theories T and T* are mutually model-consistent. Moreover, the theory T* is model-complete by Robinson's model-completeness test (Theorem 2.2). Thus, the theory $\mathcal{TA}(\mathcal{E}_T)$ is a model-companion for T.

A formulation of the notions of model-consistency and model-completeness for classes of structures will be useful later. Let Σ

be a class of similar structures. A subclass $\mathcal{C}$ of Σ is said to be *model-consistent with* Σ if each member of Σ is a substructure of a member of $\mathcal{C}$. A subclass $\mathcal{C}$ of Σ is said to be *model-complete* if whenever M and M' are members of $\mathcal{C}$ and M is a substructure of M', then M is an elementary substructure of M'.

Corollary 2.5. If $\mathcal{E}_T$ is a generalized elementary class, then $\mathcal{E}_T$ is a model-complete class.

There are two natural questions about model-companions. The first question is under what conditions is a model-companion for a theory T also a model-completion for T. Comparison of the definitions shows one requirement to be that the model-companion must contain the theory T. When this requirement is satisfied, the problem reduces to a property of the class of models of T. A theory is said to have the *amalgamation property* if whenever M, M', and M'' are models of the theory and M is a substructure of both M' and M'', then there is a model M''' of the theory such that the following diagram commutes:

$$\begin{array}{ccccc} & & M''' & & \\ & \rotatebox{45}{$\subseteq$} & & \rotatebox{-45}{$\supseteq$} & \\ M' & & & & M'' \\ & \rotatebox{-45}{$\supseteq$} & & \rotatebox{45}{$\subseteq$} & \\ & & M & & \end{array} .$$

Proposition 2.6. (Eklof & Sabbagh (34)) Assume that a theory T* is a model-companion for a theory T and that T* contains T. Then T* is a model-completion of T if and only if T has the amalgamation property.

Proof. Assume that T* is a model-completion for T. Let M, M', and M'' be models of T such that M is a substructure of both M' and M''. There are models $\overline{M'}$ and $\overline{M''}$ of $T^* \cup \mathrm{Diag}(M')$ and $T^* \cup \mathrm{Diag}(M'')$, respectively, since T* is model-consistent relative to T. Both $\overline{M'}$ and $\overline{M''}$ are models of the complete theory $T^* \cup \mathrm{Diag}(M)$, so there is a structure M''' which a a model of $T^* \cup \mathrm{Diag}(\overline{M'}) \cup \mathrm{Diag}(\overline{M''})$. Since T^* contains T, M''' is a model of T, and the following diagram commutes:

$$\begin{array}{ccccc} & & M''' & & \\ & \rotatebox{45}{$\subseteq$} & & \rotatebox{-45}{$\supseteq$} & \\ \overline{M'} & & & & \overline{M''} \\ \rotatebox{90}{$\subseteq$} & & & & \rotatebox{90}{$\subseteq$} \\ M' & & & & M'' \\ & \rotatebox{-45}{$\supseteq$} & & \rotatebox{45}{$\subseteq$} & \\ & & M & & \end{array} .$$

Conversely, assume T has the amalgamation property. Let M be a model of T, and let M' and M'' be models of T^* which extend M. The structures M' and M'' are models of T also, for T^* contains T. Since T has the amalgamation property, there is a model M''' of T such that the following diagram commutes:

$$\begin{array}{ccccc} & & M''' & & \\ & \rotatebox{45}{$\subseteq$} & & \rotatebox{-45}{$\supseteq$} & \\ M' & & & & M'' \\ & \rotatebox{-45}{$\supseteq$} & & \rotatebox{45}{$\subseteq$} & \\ & & M & & \end{array} .$$

Since T^* is model-consistent relative to T, there is an extension M'''' of M''' which is a model of T^*. The model-completeness of T^* implies that

$$\begin{array}{ccccc} & & M'''' & & \\ & \rotatebox{45}{$\prec$} & & \rotatebox{-45}{$\succ$} & \\ M' & & & & M'' \\ & \rotatebox{-45}{$\supseteq$} & & \rotatebox{45}{$\subseteq$} & \\ & & M & & \end{array} ,$$

so M' and M'' are elementarily equivalent in the language of M.

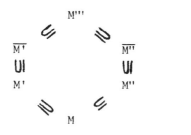

Corollary 2.7. Assume that T is an inductive theory and that T* is a model-companion for T. Then T* is a model-completion for T if and only if T has the amalgamation property.

The second question concerning model-companions is under what circumstances will a model-companion for a theory T be a complete theory. The answer involves another property of the class of models of T. A theory has the joint embedding property if for each pair of models M and M' of the theory there is a model M'' of the theory such that

$$\begin{array}{ccc} & M'' & \\ \rotatebox{45}{$\subseteq$} & & \rotatebox{-45}{$\supseteq$} \\ M & & M' \end{array} .$$

Proposition 2.8. A model-companion T* for a theory T is a complete theory if and only if T has the joint embedding property.

Proof. Assume T* is a complete theory. Let M and M' be models of T. There are models $\overline{M}$ and $\overline{M'}$ of T* which extend M and M', respectively. Since T* is complete, there is a model M'' of T* which extends both $\overline{M}$ and $\overline{M'}$. Since T is model-consistent relative to T*, there is a model M''' of T extending M''. Thus,

$$\begin{array}{ccc} & M''' & \\ & \cup\!\! \mid & \\ & M'' & \\ \rotatebox{45}{$\subseteq$} & & \rotatebox{-45}{$\supseteq$} \\ \overline{M} & & \overline{M'} \\ \cup\!\! \mid & & \cup\!\! \mid \\ M & & M' \end{array} ,$$

and T has the joint embedding property.

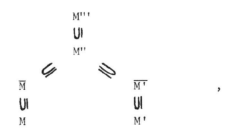

Conversely, assume T has the joint embedding property. Let M and M'' be models of T*. There are models $\overline{M}$ and $\overline{M'}$ of T which extend M and M', respectively. Since T has the joint embedding property, there is a model M'' of T extending both $\overline{M}$ and $\overline{M'}$. The model-consistency of T^* relative to T implies the existence of a model M''' of T^* extending M''. Diagrammatically again,

$$\begin{array}{ccccc} & & M''' & & \\ & & \rotatebox{90}{$\subseteq$} & & \\ & & M'' & & \\ & \rotatebox{45}{$\subseteq$} & & \rotatebox{135}{$\subseteq$} & \\ \overline{M} & & & & \overline{M'} \\ \rotatebox{90}{$\subseteq$} & & & & \rotatebox{90}{$\subseteq$} \\ M & & & & M' \end{array} .$$

Since T^* is model-complete,

$$\begin{array}{ccccc} & & M''' & & \\ & \rotatebox{45}{$\prec$} & & \rotatebox{135}{$\prec$} & \\ M & & & & M' \end{array} .$$

Thus, any two models of T^* are elementarily equivalent, and T^* is complete.

Some theories fail even to have a model-companion. The most important examples are groups (Eklof & Sabbagh (34)), division rings (Boffa & van Pragg (11), Macintyre (58), Sabbagh (91), Wheeler (107)), commutative rings with nilpotent elements (Cherlin (17)), number theory (Barwise & A. Robinson (6)), and metabelian groups (Saracino (96)). The non-existence of a model-companion means that the class of existentially complete structures is not axiomatizable. In such a situation, one must seek material analogues rather than formal analogues for algebraically closed fields. In other words one must seek a model-complete class of structures which is also model-consistent, or at least whose theory is model-consistent relative to the original theory. The techniques for obtaining such classes are finite and infinite forcing.

On the formal side, each type of forcing defines a forcing

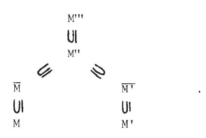

operator * which maps first order theories to companion first order theories. The significance of these new companion theories is not yet understood. However, these new forcing companions do share the following properties with model-completions and model-companions:

(1) $T^* = (T_\forall)^*$;

(2) $(T^*)_\forall = T_\forall$;

(3) $T^{**} = T^*$;

(4) $(T_1)_\forall = (T_2)_\forall$ if and only if $T_1{}^* = T_2{}^*$;

(5) T^* contains $\mathfrak{Th}(\mathcal{E}_T)$;

(6) T has the joint embedding property if and only if T^* is complete;

(7) if T has a model-companion T_1, then T^* and T_1 are logically equivalent.

CHAPTER 3

INFINITE FORCING IN MODEL THEORY

Forcing was adapted from set theory to model theory by A. Robinson in 1969 and 1970. Robinson developed two types of forcing for model theory - finite forcing and infinite forcing. To apply either type of forcing, one begins with a theory and its models. Then one constructs another class of models through forcing. These new models and their theory are analogues of algebraically closed fields and their theory. One obtains through infinite forcing a class of models which is model-complete and model-consistent relative to the original class, but the construction is highly non-effective. On the other hand, finite forcing leads only to a model-complete class whose theory is model-consistent with the original theory, but the construction is more nearly effective than that of infinite forcing.

Infinite forcing is developed in this chapter. After the definitions and basic lemmas, the class of infinitely generic structures is characterized as the unique maximal model-complete and model-consistent subclass of an inductive class. Next, it is shown that the forcing relation is completely determined by existential types. Then the relation between infinite forcing and model-companions is discussed. The chapter concludes with some remarks on the interrelationships of the various subclasses of structures which have been introduced.

§ 1 Infinite Forcing

As previously, let Σ be an inductive class of similar structures, and let $\mathcal{L}(\Sigma)$ be a language appropriate for Σ. All structures will be assumed to be members of Σ unless otherwise specified.

The relation *M infinitely forces* ϕ, denoted by $M \Vdash \phi$, between a structure M in Σ and a sentence ϕ defined in M is defined inductively as follows:

(i) if ϕ is atomic, then $M \Vdash \phi$ if and only if $M \models \phi$;

(ii) if $\phi = \psi \vee \chi$, then $M \Vdash \phi$ if and only if $M \Vdash \psi$ or $M \Vdash \chi$;

(iii) if $\phi = \psi \wedge \chi$, then $M \Vdash \phi$ if and only if $M \Vdash \psi$ or $M \Vdash \chi$;

(iv) if $\phi = \exists v_i\ \psi(v_i)$, then $M \Vdash \phi$ if and only if $M \Vdash \psi(a)$ for some element a of M;

(v) if $\phi = \sim \psi$, then $M \Vdash \phi$ if and only if there is no extension M' of M in Σ such that $M' \Vdash \psi$.

Lemma 3.1. (1) If ϕ is a sentence defined in a structure M, then M cannot infinitely force both ϕ and $\sim\phi$.

(2) If ϕ is a sentence defined in M, M' is an extension of M in Σ, and M infinitely forces ϕ, then M' infinitely forces ϕ.

Proof. Part (1) follows from the fifth clause in the definition of infinite forcing. Part (2) follows from the definition by a straightforward induction on the complexity of ϕ, which will be omitted here.

Part (2) of the lemma indicates a similarity between the forcing relation and the satisfaction relation. However, it is generally not the case that a structure will infinitely force every sentence or its negation. A structure M in Σ will be called *infinitely generic* (in Σ) if for each sentence ϕ defined in M, either $M \Vdash \phi$ or $M \Vdash \sim\phi$.

Proposition 3.2. Every structure in Σ is a substructure of an infinitely generic structure in Σ.

Proof. (The proof is similar to that of Proposition 1.3.) Let M be a member of Σ. Let $\kappa = \max\{\mathrm{card}(M), \mathrm{card}(\mathcal{L}(\Sigma)), \aleph_0\}$, and let $\{\phi_\alpha : \alpha < \kappa\}$ be an enumeration of all the sentences defined in M. Let $M_0 = M$. Construct a chain of structures from Σ as follows:

(i) $\alpha = \beta + 1$ is a successor ordinal. If $M_\beta \Vdash \phi_\beta$ or $M_\beta \Vdash \sim\phi_\beta$, then let $M_\alpha = M_\beta$. Otherwise, there is an extension M' of M_β in Σ such that $M' \Vdash \phi_\beta$. Let $M_\alpha = M'$.

(ii) α is a limit ordinal. Let $M_\alpha = \bigcup_{\beta<\alpha} M_\beta$.

Now let $M^1 = \bigcup_{\alpha<\kappa} M_\alpha$. The structure M^1 is in Σ, since Σ is an inductive class. For each sentence ϕ defined in M, either $M^1 \Vdash$ or $M^1 \Vdash \sim\phi$. Continue this construction, obtaining a chain

$$M = M^0 \subseteq M^1 \subseteq M^2 \subseteq \ldots$$

of structures in Σ such that if ϕ is a sentence defined in M^n, then either $M^{n+1} \Vdash \phi$ or $M^{n+1} \Vdash \sim\phi$. Now let $M^\omega = \bigcup_{n<\omega} M^n$. Then M^ω is an extension of M in Σ, and M^ω is infinitely generic in Σ.

The subclass of all infinitely generic structures of Σ will be denoted by $\mathcal{G}_\Sigma$. For a first order theory T, set $\Sigma = \mathcal{M}(T_\forall)$, and denote the subclass of infinitely generic structures of Σ by $\mathcal{G}_T$.

<u>Theorem 3.3</u>. A member M of Σ is infinitely generic if and only if for each sentence ϕ defined in M

$$M \models \phi \text{ if and only if } M \Vdash \phi.$$

<u>Proof</u>. Assume the second condition, i.e., that the satisfaction relation and the infinite forcing relation coincide for M. If ϕ is a sentence defined in M, then M satisfies either ϕ or $\sim\phi$, so M infinitely forces either ϕ or $\sim\phi$. Therefore, M is infinitely generic.

Conversely, assume that M is infinitely generic. Proceed by induction on the complexity of ϕ.

(i) If ϕ is atomic, then by definition $M \models \phi$ if and only if $M \Vdash \phi$.

(ii) If $\phi = \psi \vee \chi$, then the following are equivalent:

(a) $M \models \phi$,

(b) $M \models \psi$ or $M \models \chi$,

(c) $M \Vdash \psi$ or $M \Vdash \chi$ (by induction),

(d) $M \Vdash \phi$.

(iii) If $\phi = \psi \wedge \chi$, then the following are equivalent:

(a) $M \models \phi$,

(b) $M \models \psi$ and $M \models \chi$,

(c) $M \Vdash \psi$ and $M \Vdash \chi$ (by induction),

(d) $M \Vdash \phi$.

(iv) If $\phi = \exists v_i \; \psi(v_i)$, then the following are equivalent:

(a) $M \models \phi$,

(b) $M \models \psi(a)$ for some element a of M,

(c) $M \Vdash \psi(a)$ for some element a of M (by induction),

(d) $M \Vdash \phi$.

(v) If $\phi = \sim\psi$, then the following are equivalent:

(a) $M \models \phi$,

(b) not $M \models \psi$,

(c) not $M \Vdash \psi$ (by induction),

(d) $M \Vdash \phi$ (since M is infinitely generic).

Thus, the infinitely generic structures are precisely those for which the infinite forcing relation and the satisfaction relation coincide.

Proposition 3.4. If an infinitely generic structure is a substructure of a second infinitely generic structure, then the first is an elementary substructure of the second.

Proof. Let M and M' be infinitely generic structures in Σ, and suppose that M is a substructure of M'. Let ϕ be a sentence defined in M. If M satisfies ϕ, then M infinitely forces ϕ, so M' infinitely forces ϕ. Therefore, M' satisfies ϕ.

Proposition 3.5. Every infinitely generic structure in Σ is existentially complete in Σ.

Proof. Let M be an infinitely generic structure in Σ, and let ϕ be an existential sentence defined in M and true in some extension M' of M in Σ. There is an infinitely generic structure M'' of Σ extending M' (Proposition 3.2). Since ϕ is an existential sentence and is true in M', ϕ is true in M''. According to the preceding proposition, M is an elementary substructure of M'', so ϕ is true in M.

Proposition 3.6. The class $\mathcal{D}_\Sigma$ of infinitely generic structures of Σ is inductive.

Proof. Let $\{M_\alpha : \alpha < \lambda\}$ be a chain of infinitely generic structures of Σ, $M_\alpha \subseteq M_\beta$ for $\alpha < \beta$. Let $M = \bigcup_{\alpha<\lambda} M_\alpha$. Since Σ is inductive, M is in Σ. Let ϕ be a sentence defined in M. As ϕ mentions only finitely many elements of M, ϕ is defined in M_α for some $\alpha < \lambda$. Since M_α is infinitely generic, $M_\alpha \Vdash \phi$ or $M_\alpha \Vdash \sim\phi$. According to Lemma 3.1, $M \Vdash \phi$ or $M \Vdash \sim\phi$, so M is infinitely generic.

The two fundamental results on infinite forcing are the global description of the class of infinitely generic structures and the Reduction Theorem. The global description of $\mathcal{D}_\Sigma$ characterizes $\mathcal{D}_\Sigma$ without reference to forcing at all. The Reduction Theorem asserts that the forcing relation between a structure and a sentence defined in it depends only upon the existential types realized in the structure. Robinson originally used the Reduction Theorem to obtain the global description of $\mathcal{D}_\Sigma$. Subsequently, more direct proofs were found. The proof below follows an idea of G. Cherlin. The crux of the proof is the next lemma.

Lemma 3.7. A structure M is infinitely generic if and only if for each sentence of the form $\sim\psi$ which is defined in M,

$$M \models \sim\psi \text{ if and only if } M \Vdash \sim\psi.$$

Proof. The necessity of the second condition follows from Theorem 3.3. To prove sufficiency, assume that for every sentence $\sim\psi$ defined in M, $M \models \sim\psi$ if and only if $M \Vdash \sim\psi$. The proof consists of showing that for each sentence ϕ defined in M, $M \models \phi$ if and only if $M \Vdash \phi$. Proceed by induction on the complexity of ϕ.

(i) If ϕ is atomic, then by definition $M \models \phi$ if and only if $M \Vdash \phi$.

(ii) If $\phi = \psi \vee \chi$, then the following are equivalent:

(a) $M \models \phi$,

(b) $M \models \psi$ or $M \models \chi$,

(c) $M \Vdash \psi$ or $M \Vdash \chi$ (by induction),

(d) $M \Vdash \phi$.

(iii) If $\phi = \psi \wedge \chi$, then the following are equivalent:

(a) $M \models \phi$,

(b) $M \models \psi$ and $M \models \chi$,

(c) $M \Vdash \psi$ and $M \Vdash \chi$ (by induction),

(d) $M \Vdash \phi$.

(iv) If $\phi = \exists v_i\ \psi(v_i)$, then the following are equivalent:

(a) $M \models \phi$,

(b) $M \models \psi(a)$ for some element a of M,

(c) $M \Vdash \psi(a)$ for some element a of M (by induction),

(d) $M \Vdash \phi$.

(v) If $\phi = \sim\psi$, then by hypothesis $M \models \phi$ if and only if $M \Vdash \phi$.

Hence, M is infinitely generic.

Theorem 3.8. Assume Σ is an inductive class. Then the class $\mathcal{G}_\Sigma$ of infinitely generic structures is the unique subclass $\mathcal{C}$ of Σ satisfying

(1) $\mathcal{C}$ is model-consistent with Σ;

(2) $\mathcal{C}$ is model-complete;

(3) $\mathcal{C}$ contains every other subclass of Σ which satisfies conditions (1) and (2).

Proof. That $\mathcal{G}_\Sigma$ satisfies (1) and (2) follows from Proposition 3.2 and Proposition 3.4, respectively. In order to prove that $\mathcal{G}_\Sigma$ satisfies (3), suppose that $\mathcal{D}$ is a model-consistent, model-complete subclass of Σ, and let M be a member of $\mathcal{D}$. Let $\sim\psi$ be a sentence defined in M. Suppose that there is an extension M' of M in Σ which forces ψ. Since $\mathcal{G}_\Sigma$ is model-consistent with Σ, there is a structure M_1 in $\mathcal{G}_\Sigma$ which extends M'. According to Lemma 3.1, M_1 forces ψ, so M_1 satisfies ψ. Since $\mathcal{D}$ is model-consistent with Σ, there is a structure M_2 in $\mathcal{D}$ which extends M_1. Since $\mathcal{G}_\Sigma$ is model-consistent with Σ, there is a structure M_3 in $\mathcal{G}_\Sigma$ which extends M_2. Continuing in this manner, one obtains a chain $M = M_0 \subseteq M_1 \subseteq M_2 \subseteq M_3 \subseteq \ldots$ of structures such that M_{2n} is in $\mathcal{D}$ and M_{2n+1} is in $\mathcal{G}_\Sigma$ for each $n < \omega$. Since $\mathcal{D}$ and $\mathcal{G}_\Sigma$ are model-complete classes, $M_{2n} \prec M_{2n+2}$ and $M_{2n+1} \prec M_{2n+3}$ for each $n < \omega$. Let $M_\omega = \bigcup_{n<\omega} M_{2n} = \bigcup_{n<\omega} M_{2n+1}$. Then

$$\begin{array}{ccc} M = M_0 & \prec & M_\omega \\ & \subseteq \searrow \quad \nearrow \prec & \\ & M_1 & \end{array} .$$

Therefore, ψ is true in $M = M_0$, i.e., if M does not force $\sim\psi$, then M does not satisfy $\sim\psi$.

Suppose, on the other hand, that M does force $\sim\psi$. Since $\mathcal{G}_\Sigma$ is model-consistent with Σ, there is a structure M' in $\mathcal{G}_\Sigma$

which extends M. The structure M' must force $\sim\psi$, so M' satisfies $\sim\psi$. As before, one can find a member M'' of Σ such that

$$\begin{array}{ccc} M & \prec & M'' \\ & \subseteq \quad \prec & \\ & M' & \end{array} .$$

Therefore, $\sim\psi$ is true in M; so, if M forces $\sim\psi$, then M satisfies $\sim\psi$. Hence, M is infinitely generic, and $\mathcal{C}$ is contained in $\mathcal{G}_\Sigma$.

The maximality condition in the last theorem can be replaced by a closure condition on elementary substructures. The closure condition has two forms, one for arbitrary inductive classes and a simpler one for inductive, generalized elementary classes.

Theorem 3.9. Assume Σ is an inductive class. The class $\mathcal{G}_\Sigma$ of infinitely generic structures is the unique subclass $\mathcal{C}$ of Σ satisfying

(1) $\mathcal{C}$ is model-consistent with Σ;

(2) $\mathcal{C}$ is model-complete;

(3) if M is a member of Σ and M is an elementary substructure of every extension M' of M in $\mathcal{C}$, then M is in $\mathcal{C}$.

Proof. Any subclass of Σ which satisfies (1) and (2) is a subclass of $\mathcal{G}_\Sigma$, so it suffices to show that $\mathcal{G}_\Sigma$ satisfies (3) and that any class $\mathcal{D}$ satisfying (1), (2), and (3) contains $\mathcal{G}_\Sigma$.

First, suppose that M is a member of Σ which is an elementary substructure of every infinitely generic structure which extends it. Then $\mathcal{G}_\Sigma \cup \{M\}$ is a model-consistent, model-complete subclass of Σ. Consequently, $\mathcal{G}_\Sigma \cup \{M\}$ is contained in $\mathcal{G}_\Sigma$ (by the preceding theorem), and M is in $\mathcal{G}_\Sigma$. Thus, $\mathcal{G}_\Sigma$ satisfies (3).

Now suppose that $\mathcal{D}$ is a subclass of Σ which satisfies (1),

(2), and (3). Let M be a member of $\mathcal{G}_\Sigma$. Suppose that M' is a member of $\mathcal{D}$ which extends M. As in the proof of the preceding theorem, one can find an M'' in Σ such that

$$\begin{array}{ccccc} M & & \prec & & M'' \\ & \searrow\!\subseteq & & \nearrow\!\prec & \\ & & M' & & \end{array},$$

so M is an elementary substructure of M'. Since $\mathcal{D}$ satisfies (3), M is in $\mathcal{D}$. Therefore, $\mathcal{D}$ contains $\mathcal{G}_\Sigma$.

The closure condition in the preceding theorem can be simplified through the use of amalgamation when the class Σ is axiomatizable.

Theorem 3.10. Assume Σ is an inductive, generalized elementary class. The class $\mathcal{G}_\Sigma$ of infinitely generic structures is the unique subclass $\mathcal{C}$ of Σ satisfying

(1) $\mathcal{C}$ is model-consistent with Σ;

(2) $\mathcal{C}$ is model-complete;

(3) if M is a member of Σ, M' is a member of $\mathcal{C}$, and M is an elementary substructure of M', then M is in $\mathcal{C}$.

Proof. In view of the preceding theorem, it suffices to show that if $\mathcal{D}$ is a subclass of Σ satisfying (1), (2), and (3), and M is an elementary substructure of one member of $\mathcal{D}$, then M is an elementary substructure of each of its extensions in $\mathcal{D}$. So let M be a member of Σ, and let M' be a member of $\mathcal{D}$ such that M is an elementary substructure of M'. Since $\mathcal{D}$ is model-consistent with Σ and model-complete, $\mathcal{D}$ is contained in $\mathcal{G}_\Sigma$. Consequently, M' is existentially complete, and so M is existentially complete. Let M'' be another member of $\mathcal{D}$ which extends M. Because M is existentially complete, there is a member M''' of Σ such that

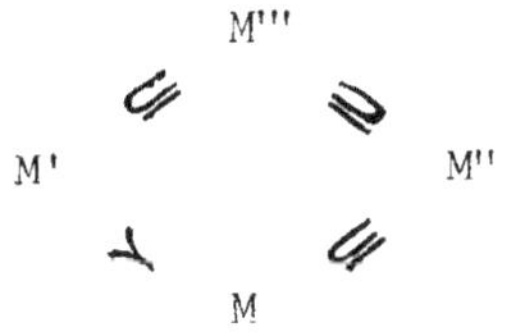

.

Since $\mathcal{D}$ is model-consistent with Σ, there is an extension M'''' of M''' in $\mathcal{D}$, and since $\mathcal{D}$ is model-complete,

.

Therefore, M is an elementary substructure of M''.

Corollary 3.10. Assume Σ is an inductive, generalized elementary class. If M is a member of Σ and $\kappa = \max\{\mathrm{card}(M), \mathrm{card}(\mathcal{L}(\Sigma)), \aleph_0\}$, then M is a substructure of an infinitely generic structure of cardinality at most κ.

Proof. The structure M is a substructure of some infinitely generic structure M'. Application of the downward Lowenheim-Skolem theorem yields a model M'' of cardinality at most κ such that $M \subseteq M'' \prec M'$. By part (3) of the preceding theorem, M'' is infinitely generic.

The Reduction Theorem asserts that the forcing relation in a generalized elementary class is determined by existential types. This reduction to the realization of existential types depends upon two facts - first, that the forcing relation is persistent under extension, and second, that structures in a generalized elementary class can be amalgamated over common existential types. These facts are used in the proof of the following lemma.

<u>Lemma 3.11</u>. Assume that Σ is an inductive, generalized elementary class. Let $\phi(v_0,\ldots, v_n)$ be a formula in the language of Σ which is either atomic or has the form $\sim\psi(v_0,\ldots, v_n)$. Suppose M and M' are two structures in Σ, and suppose $a_0, \ldots, a_n$ and $a'_0, \ldots, a'_n$ are elements of M and M', respectively, which have the same existential type in M and M', respectively, Then

$$M \Vdash \phi(a_0,\ldots, a_n) \text{ if and only if } M' \Vdash \phi(a'_0,\ldots, a'_n).$$

<u>Proof</u>. Let $\Delta(v_0,\ldots, v_n)$ be the existential type of $a_0, \ldots, a_n$ and of $a'_0, \ldots, a'_n$ in M and M', respectively. If ϕ is atomic, then the conclusion is immediate, for ϕ is either in Δ or not in Δ. Assume, then, that $\phi(v_0,\ldots, v_n)$ has the form $\sim\psi(v_0,\ldots, v_n)$, and suppose that $M \Vdash \phi(a_0,\ldots, a_n)$. Suppose further that M' does not force $\phi(a'_0,\ldots, a_n)$. Then there is an extension M'' of M' in Σ such that $M'' \Vdash \psi(a'_0,\ldots, a'_n)$. The existential type $\Delta''(v_0,\ldots, a_n)$ of $a'_0, \ldots, a'_n$ in M'' contains $\Delta(v_0,\ldots, v_n)$. According to Lemma 1.25, there is a structure M''' in Σ such that

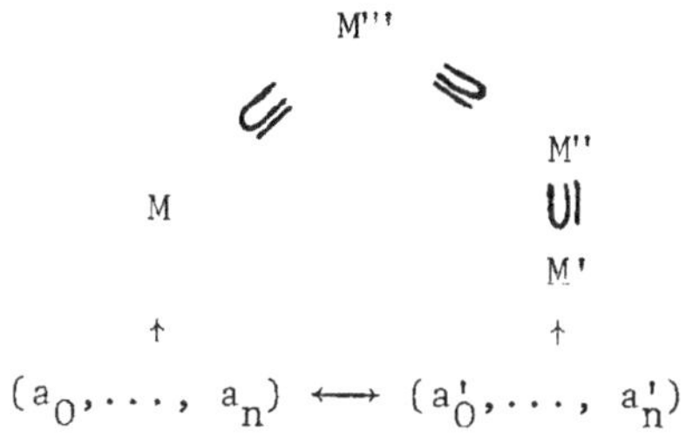

Since $M \Vdash \phi(a_0,\ldots, a_n)$, $M''' \Vdash \sim\psi(a_0,\ldots, a_n)$. Also, $M''' \Vdash \psi(a'_0,\ldots, a'_n)$, since $M'' \Vdash \psi(a'_0,\ldots, a'_n)$. But $a_0 = a'_0, \ldots,$ $a_n = a'_n$ in M''', so $M''' \Vdash \psi(a_0,\ldots, a_n)$. This contradicts that M''' cannot force both $\psi(a_0,\ldots, a_n)$ and $\sim\psi(a_0,\ldots, a_n)$.

The <u>modified</u> <u>rank</u> of a formula ϕ, denoted by m.r.(ϕ), is defined inductively as follows: if ϕ is atomic or if $\phi = \sim\psi$, then m.r.$(\phi) = 1$; if $\phi = \psi \vee \chi$ or $\phi = \psi \wedge \chi$, then m.r.(ϕ) = m.r.(ψ) + m.r.(χ); if $\phi = \exists v_i \; \psi(v_i)$, then m.r.($\phi$) = m.r.($\psi$) + 1.

The existential degree of a formula ϕ, denoted by e.d.(ϕ), is defined inductively as follows: if ϕ is atomic or if $\phi = \sim\psi$, then e.d.(ϕ) = 0; if $\phi = \psi \vee \chi$ or $\phi = \psi \wedge \chi$, then e.d.(ϕ) = e.d.(ψ) + e.d.(χ); if $\phi = \exists v_i \; \psi(v_i)$, then e.d.($\phi$) = e.d.($\psi$) + 1.

Lemma 3.12. Assume Σ is an inductive, generalized elementary class. Let $\phi(v_0, \ldots, v_n)$ be a formula in the language of Σ of modified rank one. Then there is a set R_ϕ of existential types $\Delta(v_0, \ldots, v_n)$ such that if M is a structure in Σ and $a_0, \ldots, a_n$ are elements of M, then

$M \Vdash \phi(a_0, \ldots, a_n)$ if and only if the existential type of $a_0, \ldots, a_n$ in M is in R_ϕ.

Proof. Let $R_\phi = \{\Delta(v_0, \ldots, v_n)$: there is a structure M in Σ and elements $a_0, \ldots, a_n$ in M such that $M \Vdash \phi(a_0, \ldots, a_n)$, and $\Delta(v_0, \ldots, v_n)$ is the existential type of $a_0, \ldots, a_n$ in $M\}$. The collection R_ϕ is a set because there are at most $2^{\mathrm{card}(\mathcal{L}(\Sigma))}$ existential n+1-types. If M is in Σ and $a_0, \ldots, a_n$ are elements of M such that $M \Vdash \phi(a_0, \ldots, a_n)$, then the existential type of $a_0, \ldots, a_n$ in M is in R_ϕ by definition. Conversely, if $a_0, \ldots, a_n$ are elements of a structure M in Σ, and the existential type of $a_0, \ldots, a_n$ in M is in R_ϕ, then $M \Vdash \phi(a_0, \ldots, a_n)$ according to the preceding lemma.

Theorem 3.13. (Reduction Theorem (A. Robinson)). Assume that Σ is an inductive, generalized elementary class. Let $\phi(v_0, \ldots, v_n)$ be a formula in the language of Σ, and let m be the existential degree of ϕ. Then there is a set R_ϕ of existential types $\Delta(v_0, \ldots, v_n, v_{n+1}, \ldots, v_{n+m})$ such that for each structure M in and each sequence $a_0, \ldots, a_n$ of elements of M,

$M \Vdash \phi(a_0, \ldots, a_n)$ if and only if there are elements $b_1, \ldots, b_m$ of M such that the existential type of $a_0, \ldots, a_n, b_1, \ldots, b_m$ in M is in R_ϕ.

Proof. The proof is by induction on the modified rank of ϕ. For formulas of modified rank one, this theorem coincides with the preceding lemma. So assume that $m.r.(\phi) = k > 1$ and that the theorem holds for all formulas of lower modified rank.

Suppose $\phi(v_0, \ldots, v_n) = \psi(v_0, \ldots, v_n) \vee \chi(v_0, \ldots, v_n)$. Then $m.r.(\phi) = m.r.(\psi) + m.r.(\chi)$. Let $r = e.d.(\psi)$ and $s = e.d.(\chi)$. Let $R_\psi = \{\Delta_\mu(v_0, \ldots, v_n, v_{n+1}, \ldots, v_{n+r}) : \mu < \eta\}$ and $R\chi = \{\overline{\Delta}_\nu(v_0, \ldots, v_n, v_{n+r+1}, \ldots, v_{n+r+s}) : \nu < \lambda\}$ be the sets of existential types for ψ and χ, respectively. Now let $R_\phi = \{\Delta_{\mu,\nu}(v_0, \ldots, v_n, v_{n+1}, \ldots, v_{n+r}, v_{n+r+1}, \ldots, v_{n+r+s}) : \mu < \eta,$ $\nu < \lambda\}$, where $\Delta_{\mu,\nu}(v_0, \ldots, v_n, v_{n+1}, \ldots, v_{n+r}, v_{n+r+1}, \ldots, v_{n+r+s}) =$ $\{\psi'(v_0, \ldots, v_n, v_{n+1}, \ldots, v_{n+r}) \vee \chi'(v_0, \ldots, v_n, v_{n+r+1}, \ldots, v_{n+r+s}) :$ $\psi'(v_0, \ldots, v_n, v_{n+1}, \ldots, v_{n+r}) \in \Delta_\mu$ and $\chi'(v_0, \ldots, v_n, v_{n+r+1}, \ldots, v_{n+r+s}) \in \overline{\Delta}_\nu\}$. The following are equivalent:

(1) $M \Vdash \phi(a_0, \ldots, a_n)$;

(2) $M \Vdash \psi(a_0, \ldots, a_n)$ or $M \Vdash \chi(a_0, \ldots, a_n)$;

(3) there are elements $b_1, \ldots, b_r$ of M such that the existential type of $a_0, \ldots, a_n, b_1, \ldots, b_r$ is in R_ψ or there are elements $c_1, \ldots, c_s$ of M such that the existential type of $a_0, \ldots, a_n, c_1, \ldots, c_s$ is in R_χ;

(4) there are elements $b_1, \ldots, b_r, c_1, \ldots, c_s$ of M such that the existential type of $a_0, \ldots, a_n, b_1, \ldots, b_r, c_1, \ldots, c_s$ is in R_ϕ.

Suppose $\phi(v_0, \ldots, v_n) = \psi(v_0, \ldots, v_n) \wedge \chi(v_0, \ldots, v_n)$. Let $r = e.d.(\psi)$ and $s = e.d.(\chi)$. Let

$R_\psi = \{\Delta_\mu(v_0,\ldots, v_n, v_{n+1},\ldots, v_{n+r}) : \mu < \eta\}$ and let $R_\chi = \{\overline{\Delta}_\nu(v_0,\ldots, v_n, v_{n+1+1},\ldots, v_{n+r+s}) : \nu < \lambda\}$ be the sets of existential types for ψ and χ, respectively. Now let

$R_\phi = \{\Delta_\mu(v_0,\ldots, v_n, v_{n+1},\ldots, v_{n+r}) \cup \overline{\Delta}_\nu(v_0,\ldots, v_n, v_{n+r+1},\ldots, v_{n+r+s}) :$
$\mu < \eta, \nu < \lambda\}.$

Then the following are equivalent:

(1) $M \Vdash \phi(a_0,\ldots, a_n)$;

(2) $M \Vdash \psi(a_0,\ldots, a_n)$ and $M \Vdash \chi(a_0,\ldots, a_n)$;

(3) there are elements $b_1, \ldots, b_r$ of M such that the existential type of $a_0, \ldots, a_n, b_1, \ldots, b_r$ is in R_ψ and there are elements $c_1, \ldots, c_s$ of M such that the existential type of $a_0, \ldots, a_n, c_1, \ldots, c_s$ in M is in R_χ ;

(4) there are elements $b_1, \ldots, b_r, c_1, \ldots, c_s$ of M such that the existential type of $a_0, \ldots, a_n, b_1, \ldots, b_r, c_1, \ldots, c_s$ in M is in R_ϕ.

Finally, suppose $\phi(v_0,\ldots, v_n) = \exists v_{n+1}\, \psi(v_0,\ldots, v_n, v_{n+1})$. Let $r = \text{e.d.}(\psi)$, so $\text{e.d.}(\phi) = r + 1$. Let $R_\psi = \{\Delta_\mu(v_0,\ldots, v_n, v_{n+1}, v_{n+2},\ldots, v_{n+r+1}) : \mu < \eta\}$ be the set of existential types for ψ. Let $R_\phi = R_\psi$. Then the following are equivalent:

(1) $M \Vdash \phi(a_0,\ldots, a_n)$;

(2) there is an element a of M such that $M \Vdash \psi(a_0,\ldots, a_n, a)$;

(3) there is an element a of M and that are elements $b_1, \ldots, b_r$ of M such that the existential type of $a_0, \ldots, a_n, a, b_1, \ldots, b_r$ in M is in R_ψ;

(4) there are elements $a, b_1, \ldots, b_r$ of M such that the existential type of $a_0, \ldots, a_n, a, b_1, \ldots, b_r$ in M is in R_ϕ.

For each formula ϕ in the language of Σ, the set R_ϕ of existential types constructed above is called a resultant for ϕ.

Corollary 3.14. Assume Σ is an inductive, generalized elementary class. Let $\phi(v_0,\ldots, v_n)$ be a formula in the language of Σ. Let M be an infinitely generic structure in Σ, and let $a_0, \ldots, a_n$ be elements of M. Then whether M infinitely forces $\phi(a_0,\ldots, a_n)$ depends only on the existential type of $a_0, \ldots, a_n$ in M.

Proof. The structure M satisfies $\phi(a_0,\ldots, a_n)$ if and only if it satisfies $\sim\sim\phi(a_0,\ldots, a_n)$. Since M is infinitely generic, M satisfies $\sim\sim\phi(a_0,\ldots, a_n)$ if and only if $M \Vdash \sim\sim\phi(a_0,\ldots, a_n)$. The formula $\sim\sim\phi(v_0,\ldots, v_n)$ has existential degree zero, so M satisfies $\phi(a_0,\ldots, a_n)$ if and only if the existential type of $a_0, \ldots, a_n$ in M is in $R_{\sim\sim\phi}$.

Corollary 3.15. Assume Σ is an inductive, generalized elementary class. Let M be an infinitely generic structure in Σ, and let $a_0, \ldots, a_n$ be elements of M. Then the complete type of $a_0, \ldots, a_n$ in M is uniquely determined by the existential type of $a_0, \ldots, a_n$ in M.

§ 2 Model-companions and Infinitely Generic Structures

Infinite forcing provides a generalization of the concepts of model-completion and model-companion.

Theorem 3.16. Let T be a first order theory.

(1) If T has a model-companion T^*, then $\mathcal{G}_T = \mathcal{M}(T^*)$.

(2) If $\mathcal{G}_T$ is a generalized elementary class, then $\mathcal{TA}(\mathcal{G}_T)$ is a model-companion for T.

Proof. (1) Suppose that T^* is a model-companion for T. Then $\mathfrak{M}(T^*) = \mathcal{E}_T$. Since T^* is model-complete, $\mathcal{E}_T$ is a model-complete class. Therefore, $\mathcal{E}_T$ satisfies the conditions in Theorem 3.10, so $\mathcal{G}_T = \mathcal{E}_T = \mathfrak{M}(T^*)$.

(2) Suppose that $\mathcal{G}_T$ is a generalized elementary class. Then $\mathcal{T}h(\mathcal{G}_T)$ is mutually model-consistent with respect to T and is model-complete.

Corollary 3.17. The following are equivalent for a first order theory T:

(1) T has a model-companion;

(2) $\mathcal{E}_T$ is a generalized elementary class;

(3) $\mathcal{G}_T$ is a generalized elementary class.

Moreover, if any of these conditions is satisfied, then $\mathcal{E}_T = \mathcal{G}_T$.

Thus, if a theory T does not have a model-companion, then $\mathcal{G}_T$ is not axiomatizable. Nevertheless, the class $\mathcal{G}_T$ is still the natural model-companion class for T, and its theory is of some interest. The infinite forcing companion T^F of T is defined to be the theory of $\mathcal{G}_T$.

Proposition 3.18. Let T be a first order theory. Then

(1) $T^F = (T_\forall)^F$;

(2) $(T^F)_\forall = T_\forall$;

(3) $T^{FF} = T^F$;

(4) $(T_1)_\forall = (T_2)_\forall$ if and only if $T_1{}^F = T_2{}^F$, for two first order theories T_1 and T_2;

(5) T^F is a complete theory if and only if T has the joint embedding property;

(6) $T^F \supseteq \mathcal{T}h(\mathcal{E}_T) \supseteq T_{\forall\exists}$;

(7) if T has a model-companion T^*, then T^F is the deductive closure of T^*.

Proof. Parts (1) - (4), (6), and (7) follow from either definitions or earlier results. The proof of part (5) is similar to that of Proposition 2.8.

The infinite forcing companion T^F can also be defined in terms of weak infinite forcing. A structure M in Σ is said to weakly infinitely force a sentence ϕ defined in M if no extension of M in Σ infinitely forces $\sim\phi$. Equivalently, M weakly infinitely forces ϕ if and only if $M \Vdash \sim\sim\phi$. That M weakly infinitely forces ϕ is denoted by $M \Vdash^* \phi$. A theory of weak infinite forcing paralleling that of infinite forcing can be developed. Fortunately, nothing really new occurs, as one obtains the same class $\mathcal{G}_\Sigma$ of weakly infinitely generic structures. This is the content of the following result due to C. Wood.

Proposition 3.19. Let M be a structure in Σ. Suppose that, for each sentence ϕ defined in M, if M satisfies ϕ, then $M \Vdash^* \phi$. Then M is infinitely generic.

Proof. It suffices to show that for each sentence ψ defined in M, M satisfies $\sim\psi$ if and only if $M \Vdash \sim\psi$. If M satisfies $\sim\psi$, then by hypothesis $M \Vdash^* \sim\psi$, so $M \Vdash \sim\psi$ (for if $M \Vdash \sim\sim\sim\psi$, then no extension of M in Σ can infinitely force $\sim\sim\psi$, so $M \Vdash \sim\psi$). Conversely, if $M \Vdash \sim\psi$, then M cannot force $\sim\sim\psi$, so M cannot satisfy $\sim\sim\psi$. Thus, M satisfies $\sim\psi$ if and only if $M \Vdash \sim\psi$.

In terms of weak infinite forcing, $T^F = \{\phi : \phi$ is a sentence of $\mathcal{L}(\Sigma)$ and $M \Vdash^* \phi$ for each structure M in $\Sigma\}$.

While weak infinite forcing does not generate any new generic structures, it does have one advantage over infinite forcing in that its Reduction Theorem is simpler. The Reduction Theorem for weak infinite forcing will be used in the axiomatization of $\mathcal{G}_\Sigma$ in Chapter 6.

Theorem 3.20 (Reduction Theorem for weak infinite forcing). Assume Σ is an inductive, generalized elementary class. For each formula $\phi(v_0, \ldots, v_n)$ in the language of Σ, there is a set R^*_ϕ of existential types $\Delta(v_0, \ldots, v_n)$ such that for each structure M in Σ and each sequence $a_0, \ldots, a_n$ of elements of M,

$M \Vdash^* \phi(a_0, \ldots, a_n)$ if and only if the existential type of $a_0, \ldots, a_n$ in M is in R^*_ϕ.

Proof. Let $R^*_\phi = R_{\sim\sim\phi}$ from Lemma 3.12.

§ 3 Subclasses of Σ

In this section, the class $\mathcal{P}_\Sigma$ of pregeneric structures will be defined, and the relationships between the classes $\mathcal{P}_\Sigma$, $\mathcal{E}_\Sigma$, $\mathcal{G}_\Sigma$, and $\mathcal{A}_\Sigma$ will be established.

Recall that a theory T* is the model-completion of a theory T if and only if T* contains T, T* is a model-companion for T, and whenever two models M' and M'' of T* are extensions of a model M of T, then M' and M'' are elementarily equivalent in the language of M. Of course, if T* is a model-completion for T, then the models of T* are just the infinitely generic structures for T.

A structure M in Σ is said to be pregeneric (in Σ) if whenever M' and M'' are infinitely generic structures in Σ which extend M and ϕ is a sentence defined in M, then M' satisfies ϕ if and only if M'' satisfies ϕ. The class of pregeneric structures of Σ is denoted by $\mathcal{P}_\Sigma$. The class of pregeneric structures is inductive. Moreover, since every infinitely generic structure is pregeneric, $\mathcal{P}_\Sigma$ is model-consistent with Σ.

Proposition 3.21. Assume Σ is an inductive class. Then the classes Σ, $\mathcal{P}_\Sigma$, $\mathcal{E}_\Sigma$, $\mathcal{G}_\Sigma$, and $\mathcal{A}_\Sigma$ are related as follows:

$$\begin{array}{ccccccccc} & & \mathcal{P}_\Sigma & & & & & & \\ & \rotatebox{-45}{$\supseteq$} & & \rotatebox{-45}{$\supseteq$} & & & & & \\ \Sigma & & & & \mathcal{P}_\Sigma \cap \mathcal{E}_\Sigma & \supseteq & \mathcal{G}_\Sigma & & . \\ & \rotatebox{45}{$\supseteq$} & & \rotatebox{45}{$\supseteq$} & & & & \rotatebox{45}{$\supseteq$} & \\ & & \mathcal{E}_\Sigma & \supseteq & \mathcal{A}_\Sigma & \supseteq & \mathcal{A}_\Sigma \cap \mathcal{G}_\Sigma \neq \emptyset & & \end{array}$$

Proof. Only $\mathcal{G}_\Sigma \cap \mathcal{A}_\Sigma \neq \emptyset$ needs verification. This follows because both $\mathcal{G}_\Sigma$ and $\mathcal{A}_\Sigma$ are inductive classes and are model-consistent with Σ.

One can find examples to show that each inclusion in the preceding diagram can be strict. However, when Σ is a generalized elementary class also, then the situation is more well-behaved.

Theorem 3.22. Assume Σ is an inductive, generalized elementary class. Then

(1) Every existentially complete structure is pregeneric.

(2) Every existentially universal structure is infinitely generic.

(3) The classes Σ, $\mathcal{P}_\Sigma$, $\mathcal{E}_\Sigma$, $\mathcal{G}_\Sigma$, and $\mathcal{A}_\Sigma$ are related as follows:

$$\Sigma \supseteq \mathcal{P}_\Sigma \supseteq \mathcal{E}_\Sigma \supseteq \mathcal{G}_\Sigma \supseteq \mathcal{A}_\Sigma \qquad .$$

(4) The class $\mathcal{G}_\Sigma$ of infinitely generic structures is the class of elementary substructures of members of $\mathcal{A}_\Sigma$.

Proof. (1) Since Σ is a generalized elementary class, one can amalgamate over any existentially complete structure in Σ. Suppose that M' and M'' are infinitely generic structures which extend an existentially complete structure M. Then there is a structure M''' in Σ such that

$$\begin{array}{ccccc} & & M''' & & \\ & \preceq & & \succeq & \\ M' & & & & M'' \\ & \succeq & & \preceq & \\ & & M & & \end{array} .$$

There is an infinitely generic structure M'''' extending M''' , so

$$\begin{array}{ccccc} & & M'''' & & \\ & \subseteq & & \supseteq & \\ M' & & & & M'' \\ & \succeq & & \preceq & \\ & & M & & \end{array} .$$

Therefore, if ϕ is a sentence defined in M, then M' satisfies ϕ if and only if M'' satisfies ϕ.

(2) Let M be an existentially universal structure. Since $\mathcal{G}_\Sigma$ and $\mathcal{A}_\Sigma$ are inductive classes and are model-consistent with Σ, there is an extension M' of M which is both infinitely generic and existentially universal. The structure M' is an elementary extension of M, because Σ is a generalized elementary class and M and M' are both existentially universal. Since M is an elementary substructure of an infinitely generic structure and Σ is a generalized elementary class, M is infinitely generic.

(3) This is a consequence of parts (1) and (2) and the preceding proposition.

(4) This follows from part (2), the model-consistency of $\mathcal{A}_\Sigma$ with Σ, and Theorem 3.10.

Corollary 3.23. For a first order theory T, $T^F = \mathcal{T}h(\mathcal{A}_T)$.

Corollary 3.24. If a first order theory T has a model-companion T*, then $\mathcal{E}_T = \mathcal{G}_T = \mathcal{M}(T^*)$, $T^F = \mathcal{T}h(\mathcal{E}_T) = \mathcal{T}h(\mathcal{G}_T) = \mathcal{T}h(\mathcal{A}_T)$, and T^F is the deductive closure of T*.

Again, examples show that each of the inclusions in part (3) of Theorem 3.22 can be strict. For fields, $\Sigma = \mathcal{P}_\Sigma$ is the class of integral domains; $\mathcal{E}_\Sigma = \mathcal{G}_\Sigma$ is the class of algebraically closed fields; and $\mathcal{A}_\Sigma$ is the class of universal domains. For formally real fields, $\Sigma \neq \mathcal{P}_\Sigma$, the latter being the class of integral domains whose quotient fields are formally real fields in which each element or its additive inverse is a sum of squares; $\mathcal{E}_\Sigma = \mathcal{G}_\Sigma$ is the class of real closed fields; and $\mathcal{A}_\Sigma$ is the class of real closed fields which include the completion of each of their finitely generated subfields. For R-modules, $\Sigma = \mathcal{P}_\Sigma$; $\mathcal{E}_\Sigma = \mathcal{G}_\Sigma$ (G. Sabbagh & E. Fisher); and $\mathcal{A}_\Sigma$ is the class of $\aleph_0$-homogeneous, $\aleph_0$-universal (in the sense of Jonsson) R-modules. This example has special significance, because the theory of R-modules need not have a model-companion. P. Eklof and G. Sabbagh (34) have shown that the theory of R-modules has a model-companion if and only if R is a coherent ring. Thus, if Σ is the class of R-modules of a noncoherent ring R, then $\mathcal{E}_\Sigma = \mathcal{G}_\Sigma$ even though neither is axiomatizable. Finally, for groups, $\Sigma = \mathcal{P}_\Sigma \supsetneq \mathcal{E}_\Sigma \supsetneq \mathcal{G}_\Sigma \supsetneq \mathcal{A}_\Sigma$. The class $\mathcal{G}_\Sigma$ strictly includes the class $\mathcal{A}_\Sigma$, because there are $2^{\aleph_0}$ nonisomorphic two generator groups and an existentially universal group must contain each of these, hence, must have cardinality at least $2^{\aleph_0}$. Therefore, an infinitely generic group of cardinality $\aleph_0$ cannot be existentially universal. That $\mathcal{E}_\Sigma$ strictly includes $\mathcal{G}_\Sigma$ is due to A. Macintyre (56).

CHAPTER 4

APPROXIMATING CHAINS FOR $\mathcal{G}_\Sigma$

The characterization of the class of infinitely generic structures without reference to forcing (Theorems 3.8 - 3.10) motivated efforts to find non-forcing constructions for this class. Two approaches evolved. One was to build structures which realized many existential types, for example, the existential universal structures, and then to take all elementary substructures of these existentially rich structures. This approach was developed by E. Fisher and H. Simmons. When Σ is a generalized elementary class, this method works well. However, when Σ is not a generalized elementary class, then existential types may not suffice for the determination of complete types in generic structures. There are examples in which some existentially universal structures are not infinitely generic. The second approach was the construction of approximating chains for the class of infinitely generic structures. An approximating chain is a collection $\{\mathcal{C}_n : n < \omega\}$ of subclasses of Σ such that $\mathcal{C}_0 = \Sigma$, $n < m$ implies $\mathcal{C}_n \supseteq \mathcal{C}_m$, and $\bigcap_{n<\omega} \mathcal{C}_n = \mathcal{G}_\Sigma$. G. Cherlin and D. Saracino independently discovered two different approximating chains. Subsequently, H. Simmons, J. Hirschfeld, and others introduced additional approximating chains.

Three approximating chains will be discussed in this chapter. The first will be the Cherlin chain, perhaps the most elegant of the various chains. Then the Hirschfeld chain, the "fastest" approximating chain, and a third chain, the "slowest" approximating chain, will be discussed. These latter two chains have nice analytical features, which will be the topic of Chapter 7. In this chapter, Σ will always be an inductive class of similar structures.

The Cherlin chain depends upon the notion of a persistent formula. A formula $\phi(v_0, \ldots, v_n)$ in the language of Σ is said to be $\underline{\Sigma\text{-persistent}}$ if whenever M and M' are members of Σ, M' extends M, $a_0, \ldots, a_n$ are elements of M, and M satisfies $\phi(a_0, \ldots, a_n)$, then M' satisfies $\phi(a_0, \ldots, a_n)$. In other words, $\phi(v_0, \ldots, v_n)$ is Σ-persistent if every substitution instance of ϕ in a member of Σ is persistent under extension in Σ. For example, any existential formula in the language of Σ is Σ-persistent. If Σ is a generalized elementary class, then a formula in the language of Σ is Σ-persistent if and only if it is logically equivalent in the theory of Σ to an existential formula (A. Robinson (79)).

A structure M in Σ is said to be $\underline{\Sigma\text{-persistently complete}}$ if whenever $\phi(v_0, \ldots, v_n)$ is a Σ-persistent formula, $a_0, \ldots, a_n$ are elements of M, and $\phi(a_0, \ldots, a_n)$ is true in some extension M' of M in Σ, then M satisfies $\phi(a_0, \ldots, a_n)$ also. Any Σ-persistently complete structure is also existentially complete in Σ. Moreover, if Σ is a generalized elementary class, then a structure is Σ-persistently complete if and only if it is existentially complete. Let Σ' be the subclass of Σ consisting of all the Σ-persistently complete structures in Σ.

The Cherlin chain is defined as follows. Let $\Sigma^0 = \Sigma$ and let $\Sigma^1 = \Sigma'$. Define the classes Σ^n for $n > 1$ inductively by $\Sigma^{n+1} = (\Sigma^n)'$. The chain $\Sigma = \Sigma^0 \supseteq \Sigma^1 \supseteq \Sigma^2 \supseteq \ldots$ is the Cherlin chain.

Lemma 4.1. For each $n < \omega$, Σ^n is an inductive class and is model-consistent with Σ.

Proof. The proof is by induction on n. For $n = 0$, $\Sigma^0 = \Sigma$, and the lemma is trivally true.

Let $n > 0$, and assume that Σ^{n-1} is an inductive class and is model-consistent with Σ. To show that Σ^n is model-consistent with Σ, it suffices to show that Σ^n is model-consistent with Σ^{n-1}, because by assumption Σ^{n-1} is model-consistent with Σ. Let M be a member of Σ^{n-1}, and let $\kappa = \max\{\text{card}(M), \text{card}(\mathcal{L}(\Sigma)), \aleph_0\}$. Let $\{\phi_\alpha : \alpha < \kappa\}$ be an enumeration of all the Σ^{n-1}-persistent formulas defined in M. Let $M_0 = M$. Define a chain $\{M_\alpha : \alpha < \kappa\}$ of members of Σ^{n-1} inductively as follows:

(i) $\alpha = \beta + 1$ is a successor ordinal. If M_β satisfies ϕ_β or if no extension of M_β in Σ^{n-1} satisfies ϕ_β, then let $M_\alpha = M_\beta$. Otherwise, there is an extension M' of M_β in Σ^{n-1} such that M' satisfies ϕ_β. Let $M_\alpha = M'$.

(ii) α is a limit ordinal. Let $M_\alpha = \bigcup_{\beta<\alpha} M_\beta$. M_α is in Σ^{n-1}, since Σ^{n-1} is an inductive class.

Now let $M^1 = \bigcup_{\alpha<\kappa} M_\alpha$. The structure M^1 is in Σ^{n-1}, and any Σ^{n-1}-persistent sentence defined in M and true in some extension of M^1 in Σ^{n-1} is true in M^1. Iterating this procedure yields a chain $M = M^0 \subseteq M^1 \subseteq M^2 \subseteq \ldots$ of members of Σ^{n-1} with the following property: if $\phi(v_0, \ldots, v_n)$ is a Σ^{n-1}-persistent formula, $a_0, \ldots, a_n$ are elements of M^r, and $\phi(a_0, \ldots, a_n)$ is true in some extension of M^{r+1} in Σ^{n-1}, then $\phi(a_0, \ldots, a_n)$ is true in in M^{r+1}. Let $M^\omega = \bigcup_{r<\omega} M^r$. The structure M^ω is in Σ^{n-1} and is Σ^{n-1}-persistently complete. Thus, M^ω is a member of Σ^n which extends M.

In order to show that Σ^n is inductive, let $\{M_\alpha : \alpha < \lambda\}$ be an ascending chain of members of Σ^n, where $\alpha < \beta$ implies $M_\alpha \subseteq M_\beta$. Since Σ^{n-1} contains Σ^n, $\{M_\alpha : \alpha < \lambda\}$ is also an ascending chain of members of Σ^{n-1}, so $M_\lambda = \bigcup_{\alpha<\lambda} M_\alpha$ is a member of Σ^{n-1}. It remains to show that M_λ is Σ^{n-1}-persistently complete. If λ is a successor ordinal, then $M_\lambda = M_{\lambda-1}$ is a member of Σ^n. So assume that λ is a limit ordinal. Suppose that $\phi(v_0, \ldots, v_n)$ is a Σ^{n-1}-persistent formula, that $a_0, \ldots, a_n$ are elements of M_λ, and that $\phi(a_0,\ldots, a_n)$ is true in some extension M' of M_λ in Σ^{n-1}. There is an $\alpha < \lambda$ such that $\phi(a_0,\ldots, a_n)$ is defined in M_α. Since M_α is in Σ^n, M_α is Σ^{n-1}-persistently complete. Therefore, M_α satisfies $\phi(a_0,\ldots, a_n)$. Since $\phi(v_0,\ldots, v_n)$ is Σ^{n-1}-persistent and M_λ extends M_α, M_λ satisfies $\phi(a_0,\ldots, a_n)$. Thus, M_λ is Σ^{n-1}-persistently complete and so is in Σ^n.

Lemma 4.2. Every $\exists_{n+1}$ formula is Σ^n-persistent.

Proof. This lemma depends on two observations. First, if $\phi(v_0,\ldots, v_n)$ is Σ^n-persistent, then so is $\exists v_i\ \phi(v_0,\ldots, v_i,\ldots, v_n)$. Secondly, if ϕ is Σ^n-persistent, then $\sim\phi$ is Σ^{n+1}-persistent. The lemma now follows by induction on n. Every $\exists_1$ formula is Σ^0-persistent. Assume, as the induction hypothesis, that every $\exists_{r+1}$ formula is Σ^r-persistent for some $r > 0$. Then every $\forall_{r+1}$ formula is Σ^{r+1}-persistent, so every $\exists_{r+2}$ formula is Σ^{r+1}-persistent.

Let $\Sigma^\omega = \bigcap_{n<\omega} \Sigma^n$.

Theorem 4.3. $\Sigma^\omega = \mathcal{G}_\Sigma$ for any inductive class Σ.

Proof. One must verify that Σ^ω is model-consistent with Σ, that Σ^ω is model-complete, and that Σ^ω contains every subclass $\mathcal{C}$ of Σ which is model-consistent with Σ and is model-complete.

Let M be a member of Σ. Let $M_0 = M$. Since Σ^n is model-consistent with Σ for each n, one can successively choose for each $n > 0$ a structure M_n in Σ^n such that $M_{n-1} \subseteq M_n$. Let $M_\omega = \bigcup_{n<\omega} M_n$. Then M_ω is in Σ^n for all $n \geq 0$, so M_ω is in Σ^ω.

Suppose that M and M' are members of Σ^ω and that M' extends M. Let $\phi(v_0,\ldots, v_n)$ be an $\exists_n$ formula, and let

$a_0, \ldots, a_n$ be elements of M. According to the preceding lemma, both ϕ and $\sim\phi$ are Σ^n-persistent. Since M and M' are in Σ^n, M satisfies $\phi(a_0, \ldots, a_n)$ if and only if M' satisfies $\phi(a_0, \ldots, a_n)$. Thus, M is an elementary substructure of M'.

Suppose that $\mathcal{C}$ is a subclass of Σ which is model-consistent with Σ and is model-complete. One shows by induction that $\mathcal{C}$ is a subclass of Σ^n for every n. Trivially, $\mathcal{C}$ is a subclass of $\Sigma^0 = \Sigma$. Assume that $\mathcal{C}$ is a subclass of Σ^r. Let M be a member of $\mathcal{C}$. Suppose that $\phi(v_0, \ldots, v_n)$ is a Σ^r-persistent formula, that $a_0, \ldots, a_n$ are elements of M, and that M' is a member of Σ^r which extends M and satisfies $\phi(a_0, \ldots, a_n)$. Since $\mathcal{C}$ is model-consistent with Σ, there is a member M'' of $\mathcal{C}$ which extends M'. The structure M'' is in Σ^r by the induction hypothesis. Since ϕ is Σ^r-persistent and M' satisfies $\phi(a\ , \ldots, a\ )$, M'' must satisfy $\phi(a_0, \ldots, a_n)$. M is an elementary substructure of M'', so M satisfies $\phi(a_0, \ldots, a_n)$. Thus, M is Σ^r-persistently complete, and $\mathcal{C}$ is a subclass of Σ^{r+1}. By induction, $\mathcal{C}$ is a subclass of Σ^n for all n, so $\mathcal{C}$ is a subclass of Σ^ω.

Thus, Σ^ω satisfies the three conditions of Theorem 3.8, so $\Sigma^\omega = \mathcal{G}_\Sigma$.

One should note that nowhere in the proof of the preceding theorem were the existence and properties of infinitely generic structures used. Consequently, Cherlin's method allows one to obtain all of the results of infinite forcing without the use of forcing at all.

Two other chains will be of interest in subsequent chapters. Each of these approximating chains $\{\mathcal{C}_n : n < \omega\}$ has the property that the rate of increase in the complexity of the $\mathcal{C}_n$-persistent formulas is equal to the rate of increase of the complexity of the definition of $\mathcal{C}_n$ within the analytical hierarchy. This property is the subject of Chapter 7.

The first of these chains was introduced by J. Hirschfeld. Let $\mathcal{C}$ be a class of similar structures. A structure M in $\mathcal{C}$ is said to <u>have no obstructions to elementary extension in</u> $\mathcal{C}$ if for each extension M' of M in $\mathcal{C}$ there is an extension M'' of M' in $\mathcal{C}$ such that M is an elementary substructure of M''. The subclass of $\mathcal{C}$ consisting of the members of $\mathcal{C}$ which have no obstructions to elementary extension in $\mathcal{C}$ is denoted by $\mathcal{H}(\mathcal{C})$.

Define a descending chain of subclasses of the class Σ inductively by $\mathcal{H}_0 = \Sigma$, $\mathcal{H}_{n+1} = \mathcal{H}(\mathcal{H}_n)$.

<u>Lemma 4.4</u>. (i) Each member of $\mathcal{H}_{n+1}$ is $\mathcal{H}_n$-persistently complete.

(ii) If $\phi(v_0,\ldots, v_r, v_{r+1},\ldots, v_s, v_{s+1},\ldots, v_m)$ is $\mathcal{H}_n$-persistent, then the formula $\exists v_0 \ldots \exists v_r \forall v_{r+1} \ldots \forall v_s\ \phi(v_0,\ldots, v_r, v_{r+1},\ldots, v_s, v_{s+1},\ldots, v_m)$ is $\mathcal{H}_{n+1}$-persistent.

<u>Proof</u>. (i) This is an immediate consequence of the definition.

(ii) It suffices to show that the formula $\forall v_{r+1} \ldots \forall v_s\ \phi(v_0,\ldots, v_r, v_{r+1},\ldots, v_s, v_{s+1},\ldots, v_m)$ is $\mathcal{H}_{n+1}$-persistent. Let M and M' be members of $\mathcal{H}_{n+1}$ such that M' extends M, and suppose that M satisfies $\forall v_{r+1} \ldots \forall v_s\ \phi(a_0,\ldots, a_r, v_{r+1},\ldots, v_s, a_{s+1},\ldots, a_m)$ for elements $a_0, \ldots, a_r, a_{s+1}, \ldots, a_m$ of M. Since M and M' are also members of $\mathcal{H}_n$, there is a member M'' of $\mathcal{H}_n$ such that

$$\begin{array}{ccc} M & \prec & M'' \\ & \subseteq \quad \subseteq & \\ & M' & \end{array}.$$

The sentence $\forall v_{r+1} \ldots \forall v_s\ \phi(a_0,\ldots, a_r, v_{r+1},\ldots, v_s, a_{s+1},\ldots, a_m)$ is true in M'' Let $b_{r+1}, \ldots, b_s$ be elements of M'. Then $\phi(a_0,\ldots, a_r, b_{r+1},\ldots, b_s, a_{s+1},\ldots, a_m)$ is true in M''.

Since $\phi(v_0,\ldots, v_r, v_{r+1},\ldots, v_s, v_{s+1},\ldots, v_m)$ is $\mathcal{H}_n$-persistent and M' is $\mathcal{H}_n$-persistently complete, $\phi(a_0,\ldots,a_r, b_{r+1},\ldots, b_s, a_{s+1},\ldots, a_m)$ is true in M'. As $b_{r+1}, \ldots, b_s$ were arbitrary elements of M', $\forall v_{r+1} \ldots \forall v_s\ \phi(a_0,\ldots, a_r, v_{r+1},\ldots, v_s, a_{s+1},\ldots, a_m)$ is true in M'. Thus, $\forall v_{r+1} \ldots \forall v_s\ \phi(v_0,\ldots, v_r, v_{r+1},\ldots, v_s, v_{s+1},\ldots, v_m)$ is $\mathcal{H}_{n+1}$-persistent.

<u>Theorem 4.5</u>. The following are true for any inductive class Σ:

(i) $\mathcal{G}_\Sigma \subseteq \mathcal{H}_n \subseteq \Sigma^n$ for each $n < \omega$;

(ii) $\mathcal{G}_\Sigma = \bigcap_{n<\omega} \mathcal{H}_n$.

<u>Proof</u>. (i) (By induction on n) For $n = 0$, part (i) is trivial. As the induction hypothesis, assume $\mathcal{G}_\Sigma \subseteq \mathcal{H}_r \subseteq \Sigma^r$. Let M be a member of $\mathcal{G}_\Sigma$, and let M' be a member of $\mathcal{H}_r$ which extends M. Since $\mathcal{G}_\Sigma$ is model-consistent with Σ, there is an infinitely generic structure M'' which extends M'. From the induction hypothesis M'' is in $\mathcal{H}_r$. The model-completeness of $\mathcal{G}_\Sigma$ implies that

$$\begin{array}{ccc} M & \prec & M'' \\ & \rotatebox{45}{$\subseteq$} \quad \rotatebox{-45}{$\subseteq$} & \\ & M' & \end{array}.$$

Therefore, M is in $\mathcal{H}_{r+1}$, and $\mathcal{G}_\Sigma \subseteq \mathcal{H}_{r+1}$.

It remains to show that $\mathcal{H}_{r+1}$ is contained in Σ^{r+1}. Suppose that M is in $\mathcal{H}_{r+1}$, that $\phi(v_0,\ldots, v_n)$ is a Σ^r-persistent formula, that $a_0, \ldots, a_n$ are elements of M, and that $\phi(a_0,\ldots, a_n)$ is true in some extension M' of M in Σ^r. The formula $\phi(v_0,\ldots, v_n)$ is $\mathcal{H}_r$-persistent also, because $\mathcal{H}_r \subseteq \Sigma^r$. There is an infinitely generic structure M'' extending M'. M'' is in $\mathcal{H}_r$ by the induction hypothesis. The sentence $\phi(a_0,\ldots, a_n)$ is true in M'', because M'' is in Σ^r and $\phi(v_0,\ldots, v_n)$ is

Σ^r-persistent. Since M is $\mathcal{H}_r$-persistently complete and $\phi(a_0,\ldots, a_n)$ is true in M'', $\phi(a_0,\ldots, a_n)$ is true in M. Thus, M is Σ^r-persistently complete, and $\mathcal{H}_{r+1} \subseteq \Sigma^{r+1}$.

(ii) This part is an immediate consequence of part (i) and Theorem 4.3.

The next chain is a slower approximating chain. Let $\overline{\mathcal{E}}_0 = \Sigma$. Proceed inductively. Assume that $\overline{\mathcal{E}}_n$ has been defined and that every $\exists_{n+1}$ formula in the language of Σ is $\overline{\mathcal{E}}_n$-persistent. Let F_{n+1} be a set of $\overline{\mathcal{E}}_n$-persistent formulas which includes all the $\exists_{n+1}$ formulas. If the language of Σ is countable, then we also require that F_{n+1} is a recursive set of formulas. A structure M in $\overline{\mathcal{E}}_n$ is said to be F_{n+1}-persistently complete if whenever $\phi(v_0,\ldots, v_m)$ is a formula in F_{n+1}, $a_0, \ldots, a_m$ are elements of M, and $\phi(a_0,\ldots, a_m)$ is true in some extension of M in $\overline{\mathcal{E}}_n$, then $\phi(a_0,\ldots, a_m)$ is true in M. The class $\overline{\mathcal{E}}_{n+1}$ is defined to be the subclass of $\overline{\mathcal{E}}_n$ consisting of all the F_{n+1}-persistently complete members of $\overline{\mathcal{E}}_n$. This definition is nonvacuous because of the following lemma.

Lemma 4.6. Every $\exists_{n+1}$ formula is $\overline{\mathcal{E}}_n$-persistent.

Proof. For n = 0 the lemma refers to existential formulas, and existential formulas are always persistent under extension. Let $n > 0$, and assume that the lemma is true for $n - 1$. Let $\phi(v_0,\ldots, v_m)$ be an $\exists_{n+1}$ formula, and assume that ϕ is not an $\exists_n$ formula in order to avoid trivialities. Let M and M' be members of $\overline{\mathcal{E}}_n$ such that M' extends M. Suppose that $a_0, \ldots, a_m$ are members of M for which M satisfies $\phi(a_0,\ldots, a_m)$. One may assume that ϕ has the form $\exists v_{m+1} \cdots \exists v_{m+r}\, \psi(v_0,\ldots, v_m, v_{m+1},\ldots, v_{m+r})$, where ψ is an $\forall_n$ formula. Suppose that M' does not satisfy $\phi(a_0,\ldots, a_m)$, that is, M' satisfies

$\forall v_{m+1} \cdots \forall v_{m+r} \sim\psi(a_0,\ldots, a_m, v_{m+1},\ldots, v_{m+r})$. Let $b_1, \ldots, b_r$ be elements of M for which M satisfies $\psi(a_0,\ldots, a_m, b_1,\ldots, b_r)$. The structure M' must satisfy $\sim\psi(a_0,\ldots, a_m, b_1,\ldots, b_r)$. The formula $\sim\psi(v_0,\ldots, v_m, v_{m+1},\ldots, v_{m+r})$ is logically equivalent to an $\exists_n$ formula $\psi'(v_0,\ldots, v_m, v_{m+1},\ldots, v_{m+r})$. The formula ψ' is $\overline{\mathcal{E}}_{n-1}$-persistent by the induction hypothesis and is in F_n. Since M is in $\overline{\mathcal{E}}_n$, M is F_n-persistently complete; so, M satisfies $\psi'(a_0,\ldots, a_m, b_1,\ldots, b_r)$. But then M satisfies both $\psi(a_0,\ldots, a_m, b_1,\ldots, b_r)$ and $\sim\psi(a_0,\ldots, a_m, b_1,\ldots, b_r)$, which is impossible. Hence, ϕ is $\overline{\mathcal{E}}_n$-persistent.

Theorem 4.7. The following are true for any inductive class Σ:

(i) $\mathcal{G}_\Sigma \subseteq \Sigma^n \subseteq \overline{\mathcal{E}}_n$ for all $n < \omega$;

(ii) $\mathcal{G}_\Sigma = \bigcap_{n<\omega} \overline{\mathcal{E}}_n$.

Proof. (i) For $n = 0$, $\Sigma = \Sigma^0 = \overline{\mathcal{E}}_0$. As the induction hypothesis, assume that $\Sigma^n \subseteq \overline{\mathcal{E}}_n$ for some $n \geq 0$. Every formula in F_{n+1} is $\overline{\mathcal{E}}_n$-persistent and therefore is Σ^n-persistent. Since each member of Σ^{n+1} is Σ^n-persistently complete and Σ^n is model-consistent with Σ, every member of Σ^{n+1} is F_{n+1}-persistently complete. Thus, $\Sigma^{n+1} \subseteq \overline{\mathcal{E}}_{n+1}$.

(ii) The class $\mathcal{G}_\Sigma$ is contained in $\bigcap_{n<\omega} \overline{\mathcal{E}}_n$, so $\bigcap_{n<\omega} \overline{\mathcal{E}}_n$ is model-consistent with Σ. It remains to show that this intersection is a model-complete class. Let M and M' be members of $\bigcap_{n<\omega} \overline{\mathcal{E}}_n$ such that M' is an extension of M. Let $\phi(v_0,\ldots, v_m)$ be a formula in F_n, and let $a_0, \ldots, a_m$ be elements of M. The structures M and M' are in $\overline{\mathcal{E}}_n$, and M is F_n-persistently complete. Therefore, $\phi(a_0,\ldots, a_m)$ is true in M if and only if it is true in M'. Thus, M is an elementary substructure of M', and $\mathcal{G}_\Sigma = \bigcap_{n<\omega} \overline{\mathcal{E}}_n$.

Assume, now, that Σ is an inductive, generalized elementary class. Then regardless of the choice for F_1, $\overline{\mathcal{E}}_1$ is just $\mathcal{E}_\Sigma$; so, one may as well assume that F_1 is just the set of $\exists_1$ formulas. According to Proposition 1.14, every $\exists\forall\exists$ formula is $\overline{\mathcal{E}}_1$-persistent. Consequently, we can and will require that F_n includes all the $\exists_{n+1}$ formulas for each $n > 1$. We will emphasize the presence of this new requirement by writing $\mathcal{E}_n$ for $\overline{\mathcal{E}}_n$.

Lemma 4.8. Every $\exists_{n+2}$ formula is $\mathcal{E}_n$-persistent for $n > 0$.

Proof. The proof is analogous to that of Lemma 4.6.

Theorem 4.9. Assume Σ is an inductive, generalized elementary class. Then

(i) $\mathcal{G}_\Sigma \subseteq \Sigma^n \subseteq \mathcal{E}_n$ for all $n < \omega$;

(ii) $\mathcal{G}_\Sigma = \bigcap_{n<\omega} \mathcal{E}_n$.

Proof. The proof is analogous to that of Theorem 4.7.

Furthermore, unless otherwise stated, we will assume that F_1 is just the set of $\exists_1$ formulas and that F_n is just the set of $\exists_{n+1}$ formulas for $n > 1$.

CHAPTER 5

FINITE FORCING IN MODEL THEORY

Finite forcing in model theory is more similar to Cohen's forcing in set theory than is infinite forcing. Finite forcing is a syntactic relation between finite sets of basic sentences, called conditions, and sentences. As in Cohen's forcing, finitely generic models are constructed for countable theories via complete sequences of conditions. For theories with model-companions, the finitely generic structures and the finite forcing companion coincide with the infinitely generic models and the infinite forcing companion. However, finite forcing and infinite forcing can diverge for theories without model-companions. Examples are number theory, groups, metabelian groups, division rings, and commutative rings with nilpotent elements.

This chapter consists of five sections: finite forcing and the finite forcing companion, finite forcing by structures and finitely generic structures, the relation between the finite forcing companion and the finitely generic structures, finite forcing companions and model-companions, and approximating theories for the finite forcing companion.

§ 1 Finite Forcing

Finite forcing is a syntactic relation. Let $\mathcal{L}$ be a first order language and let A be a set of constant symbols none of which occur in $\mathcal{L}$. $\mathcal{L}(A)$ will denote the expanded language with the function and relation symbols of $\mathcal{L}$ and the constant symbols of both $\mathcal{L}$ and A. $\mathcal{L}(A)$ is called a normal expansion of $\mathcal{L}$ if A is an infinite set. A basic sentence in $\mathcal{L}(A)$ is a sentence which is either atomic or the negation of an atomic sentence.

Let T be a theory in $\mathcal{L}$ and let $\mathcal{L}(A)$ be a normal expansion of $\mathcal{L}$. A condition in $\mathcal{L}(A)$ relative to T is a finite set P of basic sentences from $\mathcal{L}(A)$ such that $T \cup P$ is a consistent set of sentences in $\mathcal{L}(A)$. A condition P in $\mathcal{L}(A)$ relative to T is said to force a sentence ϕ in $\mathcal{L}(A)$, denoted by $P \Vdash_{T,A} \phi$, provided

(i) if ϕ is atomic, then $P \Vdash_{T,A} \phi$ if and only if $\phi \in P$;

(ii) if $\phi = \psi \vee \chi$, then $P \Vdash_{T,A} \phi$ if and only if $P \Vdash_{T,A} \psi$ or $P \Vdash_{T,A} \chi$;

(iii) if $\phi = \psi \wedge \chi$, then $P \Vdash_{T,A} \phi$ if and only if $P \Vdash_{T,A} \psi$ and $P \Vdash_{T,A} \chi$;

(iv) if $\phi = \exists v\, \psi(v)$, then $P \Vdash_{T,A} \phi$ if and only if $P \Vdash_{T,A} \psi(t)$ for some closed term t in $\mathcal{L}(A)$;

(v) if $\phi = \sim\psi$, then $P \Vdash_{T,A} \phi$ if and only if for all conditions Q containing P, Q does not force ϕ.

The set A of constant symbols does not play a distinguished role in finite forcing.

Lemma 5.1. Let $\mathcal{L}(A)$ and $\mathcal{L}(A')$ be two normal expansion of $\mathcal{L}$ and assume A' contains A. If T is a theory in $\mathcal{L}$, P is a condition in $\mathcal{L}(A)$ relative to T, and ϕ is a sentence in $\mathcal{L}(A)$, then

$$P \Vdash_{T,A} \phi \text{ if and only if } P \Vdash_{T,A'} \phi.$$

Proof. Whether a condition forces a sentence depends only upon finite chains of conditions of bounded length. Since A is infinite, any finite chain of conditions in A' can be duplicated in A. For details, see Robinson (76).

As was the case for existentially complete structures, existentially universal structures, and infinite forcing, the finite forcing relation is determined by the universal consequences of a theory.

Lemma 5.2. Let T be a theory in $\mathcal{L}$ and let P be a finite set of basic sentences in the normal expansion $\mathcal{L}(A)$. Then P is a condition relative to T if and only if it is a condition relative to $T_\forall$. Moreover, if ϕ is a sentence in $\mathcal{L}(A)$, then

$$P \Vdash_{T,A} \phi \text{ if and only if } P \Vdash_{T_\forall,A} \phi.$$

Proof. If $T \cup P$ is consistent, then clearly $T_\forall \cup P$ is consistent. Conversely, if $T_\forall \cup P$ is consistent, then it has a model M. Since $(T_\forall \cup P)_\forall$ contains $T_\forall$, there is a model M' of T which contains M. Then M' is a model of $T \cup P$, so $T \cup P$ is consistent. Moreover, since T and $T_\forall$ have the same set of conditions and since the forcing relation is entirely determined by the set of conditions,

$$P \Vdash_{T,A} \phi \text{ if and only if } P \Vdash_{T_\forall,A} \phi.$$

In view of Lemma 5.1, the set A of constants need not be specified explicitly. Furthermore, as the theory T is usually fixed in any particular situation, we will say merely that P is a condition which forces ϕ and will write $P \Vdash \phi$ for $P \Vdash_{T,A} \phi$.

Lemma 5.3. (i) If P is a condition and ϕ is a sentence, then P cannot force both ϕ and $\sim\phi$.

(ii) If P and Q are conditions, Q contains P, and ϕ is a sentence such that $P \Vdash \phi$, then $Q \Vdash \phi$.

(iii) If P is a condition and ϕ is a basic sentence such that $P \Vdash \phi$, then $P \cup \{\phi\}$ is a condition.

Proof. (i) and (ii) are immediate consequences of the definition of finite forcing. (iii) is trivial if ϕ is atomic. So assume $\phi = \sim\psi$ where ψ is an atomic sentence and $P \Vdash \phi$. If $P \cup \{\phi\}$ is not consistent, then $T \cup P \vdash \sim\phi$ or $T \cup P \vdash \psi$. Since $T \cup P$ is consistent, $Q = P \cup \{\psi\}$ is a condition. But then $Q \Vdash \psi$ which contradicts that $P \Vdash \phi$. Therefore $P \cup \{\phi\}$ is a condition.

The notion of weak finite forcing is more important than was the notion of weak infinite forcing. A condition P in the normal expansion $\mathcal{L}(A)$ is said to <u>weakly force</u> a sentence ϕ in $\mathcal{L}(A)$ if P forces $\sim\sim\phi$. That $P \Vdash \sim\sim\phi$ will be denoted by $P \Vdash^* \phi$. The standard lemmas about forcing hold for weak finite forcing.

<u>Lemma 5.4</u>. (i) If P is a condition and ϕ is a sentence, then not both $P \Vdash^* \phi$ and $P \Vdash^* \sim\phi$.

(ii) If $P \Vdash^* \phi$ and Q is a condition and contains P, then $Q \Vdash^* \phi$.

(iii) If $P \Vdash \phi$, then $P \Vdash^* \phi$.

(iv) If $P \Vdash^* \sim\phi$, then $P \Vdash \sim\phi$.

<u>Proof</u>. Parts (i), (ii), and (iii) follow directly from the definition of weak forcing. To prove part (iv), assume $P \Vdash^* \sim\phi$, i.e., $P \Vdash \sim\sim\sim\phi$. Suppose Q is a condition which contains P and forces ϕ. Then Q forces $\sim\sim\phi$, which contradicts that P forces $\sim\sim\sim\phi$. Therefore, no condition extending P forces ϕ, so P forces $\sim\phi$.

When P is a condition relative to T, then $T^f(P)$ will denote the set of sentences of $\mathcal{L}$ which are weakly finitely forced by P. In particular T^f will denote the set of sentences weakly finitely forced by the empty set $\emptyset$, i.e., $T^f = T^f(\emptyset)$. T^f is called the <u>finite forcing companion</u> of T.

<u>Proposition 5.5</u>. If T is a theory in $\mathcal{L}$ and P is a condition relative to T, then $T^f(P)$ is a consistent set of sentences in $\mathcal{L}$ and is closed under deduction. In particular, for any theory T, T^f is a consistent set of sentences.

<u>Proof</u>. See Barwise and Robinson (6).

Proposition 5.6. If T is a theory in $\mathcal{L}$, then $T^f = (T_\forall)^f$.

Proof. Immediate from Lemma 5.2.

Proposition 5.7. If P is a condition in $\mathcal{L}(A)$ relative to T and if ϕ is a universal sentence in $\mathcal{L}(A)$, then $P \Vdash \phi$ if and only if $T \cup P \vdash \phi$.

Proof. Since ϕ is universal, ϕ is logically equivalent to a sentence $\sim\exists v_0 \ldots \exists v_n\, \psi(v_0,\ldots, v_n)$ where ψ is quantifier-free and is in disjunctive normal form. Denote $\sim\exists v_0 \ldots \exists v_n\, \psi(v_0,\ldots, v_n)$ by ϕ'. The condition P forces ϕ if and only if it forces ϕ', because if it forces one, then it must weakly force the other and consequently must force the other sentence (since both ϕ and ϕ' are universal sentences). Now the following are equivalent:

(i) P does not force ϕ;

(ii) P does not force ϕ';

(iii) there is a condition Q extending P such that Q forces $\exists v_0 \ldots \exists v_n\, \psi(v_0,\ldots,v_n)$;

(iv) there is a condition Q extending P such that Q forces $\psi(a_0,\ldots,a_n)$ for some set of constants $a_0, \ldots, a_n$;

(v) there is a condition Q extending P which for some set of constants $a_0, \ldots, a_n$ contains all of the basic sentences in one of the conjuncts of $\psi(a_0,\ldots, a_n)$;

(vi) the set of sentences $T \cup P \cup \{\sim\phi'\}$ is consistent;

(vii) the set of sentences $T \cup P \cup \{\sim\phi\}$ is consistent.

Corollary 5.8. $T_\forall = (T^f)_\forall$.

Lemma 5.9. Suppose $P(a_0,\ldots, a_n)$ is a condition, where $a_0, \ldots, a_n$ are constants in $\mathcal{L}(A) - \mathcal{L}$, and suppose that $t_0, \ldots, t_n$ are closed terms of $\mathcal{L}(A)$ such that the set $P(t_0,\ldots, t_n)$ of sentences obtained by replacing each occurrence of a_i by an occurrence of t_i is a ndition also. If $\phi(a_0, \ldots, a_n, b_1, \ldots, b_r)$ is a sentence in $\mathcal{L}(A)$, $\phi(t_0, \ldots, t_n, b_1, \ldots, b_r)$ is the sentence obtained by replacing each ccurrence of a_i by an occurrence of t_i, and

$P(a_0, \ldots, a_n)$ forces $\phi(a_0, \ldots, a_n, b_1, \ldots, b_r)$, then $P(t_0, \ldots, t_n)$ forces $\phi(t_0, \ldots, t_n, b_1, \ldots, b_r)$.

Proof. The proof is by induction on the complexity of the formula $\phi(v_0, \ldots, v_n, v_{n+1}, \ldots, v_{n+r})$. For simplicity denote $\phi(a_0, \ldots, a_n, b_1, \ldots, b_r)$ by ϕ' and $\phi(t_0, \ldots, t_n, b_1, \ldots, b_r)$ by ϕ''. Also denote $P(a_0, \ldots, a_n)$ by P' and $P(t_0, \ldots, t_n)$ by P''.

(i) ϕ is atomic. P' forces ϕ' if and only if ϕ' is in P', n which case ϕ'' is in P'' and P'' forces ϕ''.

(ii) $\phi = \psi \vee \chi$. Each of the following clauses implies its successor: P' forces ϕ'; P' forces ψ' or P' forces χ'; P'' forces ψ'' r P'' forces χ''; P'' forces ϕ''.

(iii) $\phi = \psi \wedge \chi$. Each of the following clauses implies its successor: P' forces ϕ'; P' forces ψ' and P' forces χ'; P'' forces ψ'' nd P'' forces χ''; P'' forces ϕ''.

(iv) $\phi = \exists v\, \psi(v)$. Each of the following clauses implies its uccessor: P' forces $\exists v\, \psi(v, a_0, \ldots, a_n, b_1, \ldots, b_r)$; there is a losed term $t(a_0, \ldots, a_n, b_1, \ldots, b_r, b_{r+1}, \ldots, b_s)$ such that P' orces $\psi(t(a_0, \ldots, a_n, b_1, \ldots, b_r, b_{r+1}, \ldots, b_s), a_0, \ldots, a_n, b_1, \ldots, b_r))$ P'' forces $\psi(t(t_0, \ldots, t_n, b_1, \ldots, b_r, b_{r+1}, \ldots, b_s), t_0, \ldots, t_n, b_1, \ldots b_r))$ P'' forces $\exists v\, \psi(v, t_0, \ldots, t_n, b_1, \ldots, b_r)$.

(v) $\phi = \sim\psi$. Suppose P' forces ϕ' but P'' does not force ϕ''. hen there is a condition Q which extends P'' and forces ψ''. Let $\bar{a}_0, \ldots, \bar{a}_n$ be new constants not in $\mathcal{L}(A)$ and let $P(\bar{a}_0, \ldots, \bar{a}_n)$ be btained by replacing each occurrence of a_i in P' by an occurrence f $\bar{a}_i$. Then $P(\bar{a}_0, \ldots, \bar{a}_n)$ forces $\phi(\bar{a}_0, \ldots, \bar{a}_n, b_1, \ldots, b_r)$. The set $= Q \cup P(\bar{a}_0, \ldots, \bar{a}_n) \cup \{\bar{a}_0 = t_0, \bar{a}_1 = t_1, \ldots, \bar{a}_i = t_i, \ldots, \bar{a}_n = t_n\}$

is a condition because Q contains P". Then $\overline{Q}$ forces $\psi(t_0,\ldots, t_n, b_1,\ldots, b_r)$, since it contains Q; and $\overline{Q}$ forces $\sim\psi(\overline{a}_0,\ldots, \overline{a}_n, b_1,\ldots, b_r)$, since it contains $P(\overline{a}_0,\ldots, \overline{a}_n)$. Thus $\overline{Q}$ forces $\exists v_0 \ldots \exists v_n \exists v_{n+1} \ldots \exists v_{2n+1} (\psi(v_0,\ldots, v_n, b_1,\ldots, \wedge\sim\psi(v_{n+1},\ldots, v_{2n+1}, b_1,\ldots, b_r) \wedge v_0 = v_{n+1} \wedge v_1 = v_{n+2} \wedge \ldots \wedge v_n = v_{2n+1})$. But this last sentence is always invalid, which contradicts that $T^f(\overline{Q})$ is a consistent set of sentences.

The finite forcing companion contains axioms for all instances in which the weak finite forcing relation holds. This fact, as formulated in the next proposition, will be crucial in the characterizations of T^f and the finitely generic models without reference to finite forcing.

<u>Proposition 5.10</u>. Let $P(a_0,\ldots, a_n)$ be a condition in $\mathcal{L}(A)$ relative to T and let $\phi(a_0,\ldots, a_n)$ be a sentence in $\mathcal{L}(A)$, where $a_0,\ldots, a_n$ are all of the constants from $\mathcal{L}(A) - \mathcal{L}$ which occur in either P or ϕ. If $P \Vdash^* \phi$, then T^f contains the sentence

$$\forall v_0 \ldots \forall v_n (\bigwedge P(v_0,\ldots, v_n) \rightarrow \phi(v_0,\ldots, v_n)),$$

where $P(v_0,\ldots, v_n)$ and $\phi(v_0,\ldots, v_n)$ are obtained from $P(a_0,\ldots, a_n)$ and $\phi(a_0,\ldots, a_n)$ by replacing each occurrence of a_i by v_i, and $\bigwedge P(v_0,\ldots, v_n)$ denotes the conjunction of all the formulas in $P(v_0,\ldots, v_n)$.

<u>Proof</u>. If $\forall v_0 \ldots \forall v_n (\bigwedge P(v_0,\ldots, v_n) \rightarrow \phi(v_0,\ldots, v_n))$ is not in T^f, then there is a condition Q which forces $\exists v_0 \ldots \exists v_n(\bigwedge P(v_0,\ldots, v_n) \wedge \sim\phi(v_0,\ldots, v_n))$, i.e., for some closed terms $t_0, \ldots, t_n$, Q forces $\bigwedge P(t_0,\ldots, t_n)$ and Q forces $\sim\phi(t_0,\ldots, t_n)$. By lemma 5.3, $Q \cup P(t_0,\ldots, t_n)$ is a condition, and $Q \cup P(t_0,\ldots, t_n)$ forces $\sim\phi(t_0,\ldots, t_n)$. But this contradicts the preceding lemma.

§ 2 Finite Forcing by Structures

Let T be a theory in a language $\mathcal{L}$, and let M be a member of Σ_T. Let A be an infinite set of new constants such that card(A) $\geq$ card(M). Assign names from A to a set of generators of M so that every member of M is represented by a closed term of $\mathcal{L}(A)$. Let these be the names used in Diag(M). That M is in Σ_T is equivalent to the assertion that each finite subset of Diag(M) is a condition in $\mathcal{L}(A)$ relative to T.

The structure M is said to finitely force a sentence ϕ in $\mathcal{L}(A)$ if there is a finite subset P of Diag(M) such that P forces ϕ. "M finitely forces ϕ " will be denoted by " $M \Vdash \phi$ ". The finite forcing relation between a structure and sentences defined in it is independent of both the set A of constants and the assignment of names to a set of generators of M. Specifically, if $\phi(v_0, \ldots, v_n)$ is a formula of $\mathcal{L}$, and if $t_0, \ldots, t_n$ and $t_0', \ldots, t_n'$ are closed terms in two normal expansions $\mathcal{L}(A)$ and $\mathcal{L}(A')$ which name the same elements of M, respectively, then $M \Vdash \phi(t_0, \ldots, t_n)$ in the language $\mathcal{L}(A)$ if and only if $M \Vdash \phi(t_0', \ldots, t_n')$ in the language $\mathcal{L}(A')$ (see Barwise and Robinson (6)).

As in infinite forcing, structures for which finite forcing and satisfaction coincide are analogues of algebraically closed fields. A structure M is said to be finitely generic for a theory T if M is a member of Σ_T and for every sentence ϕ defined in M, M satisfies ϕ if and only if M finitely forces ϕ.

The existence of finitely generic structures for a theory is a far more delicate problem than the existence of infinitely generic structures. For infinite forcing, not only do infinitely generic structures always exist, but moreover every model is contained in an infinitely generic structure. In contrast, S. Shelah and P. Henrard have given examples of uncountable theories which have no finitely generic structures at all.

For countable theories, however, finitely generic structures always do exist.

Theorem 5.11. Let T be a theory in a countable language $\mathcal{L}$ and let $\mathcal{L}(A)$ be a normal expansion of $\mathcal{L}$, where A is a countable set. Let P be any condition in $\mathcal{L}(A)$ relative to T. Then there is a finitely generic structure M for T with names from $\mathcal{L}(A)$ such that P is contained in the diagram of M.

Proof. The proof consists of three parts: first, the construction of an ascending chain of conditions; secondly, the use of this chain to build a structure for $\mathcal{L}$; and thirdly, the demonstration that this structure is finitely generic for T.

Let $\{\phi_n : n < \omega\}$ be an enumeration of all the sentences of $\mathcal{L}(A)$. Let $P_0 = P$. Define a chain of conditions inductively as follows. Assume P_n has been obtained. If $P_n \Vdash \phi_n$ or if $P_n \Vdash \sim\phi_n$, then let $P_{n+1} = P_n$. Otherwise, there is a condition Q containing P_n such that $Q \Vdash \phi_n$. Let $P_{n+1} = Q$. Now let $S = \{P_n : n < \omega\}$. S is an ascending chain of conditions with the property that for each sentence ϕ in $\mathcal{L}(A)$ there is an $n < \omega$ such that $P_n \Vdash \phi$ or $P_n \Vdash \sim\phi$.

A finitely generic structure is constructed from the closed terms of $\mathcal{L}(A)$ using S. Let C be the set of closed terms of $\mathcal{L}(A)$. Define a binary relation $\sim$ on C by setting $t \sim t'$ for closed terms t and t' in C if and only if there is an $n < \omega$ such that $P_n \Vdash t = t'$, where P_n is the $n+1^{\underline{st}}$ member of S. The relation $\sim$ is an congruence relation on C. Denote the equivalence class of a closed term t in C by $\bar{t}$. Let M be the set of equivalence classes of C with respect to $\sim$. If $f(v_0, \ldots, v_n)$ is an $n+1$-ary function, define $f(\bar{t}_0, \ldots, \bar{t}_n) = \overline{f(t_0, \ldots, t_n)}$. The functions are well-defined, because $\sim$ is a congruence relation on C. If $R(v_0, \ldots, v_n)$ is an $n+1$-ary relation and $\bar{t}_0, \ldots, \bar{t}_n$ are members of M, define $R(\bar{t}_0, \ldots, \bar{t}_n)$ to hold in M if and only if there is an

$n < \omega$ such that $P_n \Vdash R(t_0, \ldots, t_n)$. Suppose $t_i \sim t_i'$, $0 \leq i \leq n$, and $P_m \Vdash R(t_0, \ldots, t_n)$ for some $m < \omega$. Then there is an $r \geq m$ such that $P_r \Vdash t_0 = t_0', \ldots, P_r \Vdash t_n = t_n'$, and $P_r \Vdash R(t_0, \ldots, t_n)$. Then no condition P_s containing P_r can force $\sim R(t_0', \ldots, t_n')$, so there is an $s \geq r$ such that $P_s \Vdash R(t_0', \ldots, t_n')$. Thus, the relations in M are well-defined. If c is a constant in $\mathcal{L}(A)$, assign c to name the class $\bar{c}$. M is now a structure for $\mathcal{L}(A)$, and every member of M is named by some closed term of $\mathcal{L}(A)$.

That M is finitely generic remains to be shown. The definition of functions and relations in M ensures that a basic sentence defined in M holds in M if and only if it is forced by one of the conditions in S. Let D be a finite subset of the diagram of M in the language of $\mathcal{L}(A)$. Then each sentence in D is forced by some condition in S, so there is an $n < \omega$ such that P_n forces each sentence in D. But then P_n forces the conjunction of all the sentences in D, so by Lemma 5.3.(iii), $P_n \cup D$ is a condition relative to T. Hence, M is in Σ_T. It remains to show that for every sentence ϕ in $\mathcal{L}(A)$, $M \Vdash \phi$ if and only if $M \models \phi$. Let $\mathcal{S}(M)$ be the set of sentences ϕ in $\mathcal{L}(A)$ which M forces. From the definitions of the functions and relations in M, it follows that each P_n is in the diagram of M and that each atomic sentence in $\mathcal{L}(A)$ is in $\mathcal{S}(M)$ if and only if M satisfies it. One now proceeds by induction on the complexity of sentences. Since the inductive definitions for forcing and satisfaction are identical for conjunction, disjunction, and existential quantification, one must only show that for every sentence $\sim\phi$ in $\mathcal{L}(A)$, $M \Vdash \sim\phi$ if and only if $M \models \sim\phi$. First, assume $M \Vdash \sim\phi$. Then M does not force ϕ, so by the induction hypothesis M does not satisfy ϕ. Therefore, M satisfies $\sim\phi$. Conversely, assume M satisfies $\sim\phi$. Then M does not satisfy ϕ, so M does not force ϕ. For some $n < \omega$, $\sim\phi$ is the sentence ϕ_n, so either $P_{n+1} \Vdash \phi$ or $P_{n+1} \Vdash \sim\phi$. Since P_{n+1} is a subset of

Diag(M) and M does not force ϕ, $P_{n+1} \Vdash \sim\phi$. Therefore $M \Vdash \sim\phi$. Hence, M is finitely generic

An ascending chain of conditions $\{P_n : n < \omega\}$ with $P_n \subseteq P_m$ for $n \leq m$ is called a <u>complete sequence of conditions</u> if for every sentence ϕ in $\mathcal{L}(A)$ there is an $n < \omega$ such that either $P_n \Vdash \phi$ or $P_n \Vdash \sim\phi$. The latter part of the last proof established the following corollary.

<u>Corollary 5.12</u>. A complete sequence $S = \{P_n : n < \omega\}$ of conditions determines a unique, finitely generic structure M_S.

The class of finitely generic structures for a theory T will be denoted by $\mathcal{F}_T$. $\mathcal{F}_T$ is a subclass of Σ_T of course. The basic facts about finitely generic structures are stated in the following propositions.

<u>Proposition 5.13</u>. Every finitely generic structure is existentially complete.

<u>Proof</u>. Let M be a finitely generic structure in Σ_T and let M' be an extension of M in Σ_T. Let $\phi = \exists v_0 \ldots \exists v_n\ \psi(v_0, \ldots, v_n)$ be an existential sentence defined in M, where ψ is quantifier-free. Suppose M' satisfies ϕ. Then there are elements $a_0, \ldots, a_n$ in M' such that M' satisfies $\psi(a_0, \ldots, a_n)$. Let P be the set of basic sentences occurring in $\psi(a_0, \ldots, a_n)$ which are true in M'. Then P forces $\psi(a_0, \ldots, a_n)$, so P forces ϕ. Let D be a finite subset of Diag(M). $P \cup D$ is a condition, since M' contains M, and $P \cup D$ forces ϕ since P does. Therefore M cannot force $\sim\phi$, so M cannot satisfy $\sim\phi$ and must satisfy ϕ. Hence, M is existentially complete.

<u>Proposition 5.14.</u> If M and M' are finitely generic and M' extends M, then M' is an elementary extension of M.

Proof. Let ϕ be a sentence defined and true in M. Since M is finitely generic, there is a condition P contained in Diag(M) such that $P \Vdash \phi$. Since M' contains M, P is contained in Diag(M') and P still forces ϕ. As M' is finitely generic, M' satisfies ϕ.

Proposition 5.15. An existentially complete substructure of a finitely generic structure is finitely generic also.

Proof. Let M and M' be members of Σ_T such that M' is finitely generic, M' extends M, and M is existentially complete in M'. The crux of the proof is showing that for any sentence ϕ defined in M, M forces ϕ if and only if M' forces ϕ. Clearly if M forces ϕ, then M' forces ϕ. To prove the converse, suppose ϕ is a sentence defined in M and forced by M'. Let Q be a finite subset of Diag(M') such that $Q \Vdash \phi$. Let $b_0, \ldots, b_n$ be the names of elements in M' - M which occur in Q. Let $\psi(b_0, \ldots, b_m)$ be the conjunction of the sentences in Q. Then $\exists v_0 \ldots \exists v_n \, \psi(v_0, \ldots, v_m)$ is an existential sentence defined in M and true in M', so M satisfies this sentence. Let $t_0, \ldots, t_m$ be closed terms naming elements of M for which M satisfies $\psi(t_0, \ldots, t_m)$. Let P be the set of conjuncts of $\psi(t_0, \ldots, t_m)$. Then P is a subset of Diag(M) and $P \Vdash \phi$ (Lemma 5.9). Thus $M \Vdash \phi$.

M is finitely generic provided that for each sentence ϕ defined in M, M satisfies ϕ if and only if M forces ϕ. As the set of sentences forced by M contains Diag(M) and is closed under conjunction, disjunction, and existential quantification, it suffices to show that for every sentence $\sim\phi$ defined in M, M satisfies $\sim\phi$ if and only if M forces $\sim\phi$. If M forces $\sim\phi$, then M does not force ϕ so by induction M does not satisfy ϕ and consequently M satisfies $\sim\phi$. Conversely, suppose M satisfies $\sim\phi$. Then M does not satisfy ϕ, so M does not force ϕ. Therefore M' does not force ϕ. Since M' is finitely generic, M' must force $\sim\phi$, so M forces $\sim\phi$.

Proposition 5.16. The class of finitely generic structures is an inductive class .

Proof. Let $\{M_\alpha : \alpha < \lambda\}$ be an ascending chain of finitely generic structures and let $M = \bigcup_{\alpha<\lambda} M_\alpha$. The structure M is a member of Σ_T. Since $\{M_\alpha : \alpha < \lambda\}$ is an elementary chain (Proposition 5.14), M is an elementary extension of each M_α. Let ϕ be a sentence defined in M. The sentence ϕ is defined in M_β for some $\beta < \lambda$. The following assertions are equivalent: M satisfies ϕ; M_β satisfies ϕ; M_β forces ϕ; M forces ϕ. Therefore, M is finitely generic.

§ 3 Finitely Generic Structures and the Finite Forcing Companion

The nonexistence of finitely generic structures for some theories indicates that the relation between the finite forcing companion and the finitely generic structures is substantially different from the corresponding situation for infinite forcing. With regard to infinite forcing, a theory always has infinitely generic structures and the infinite forcing companion is just the theory of these structures. With regard to finite forcing, the finite forcing companion always exists, but in Henrard's and Shelah's examples no finitely generic structures exist. So the finite forcing companion need not be the theory of some natural class of structures. The next three results clarify the relationship between the finitely generic structures and the finite forcing companion.

Proposition 5.17. Every finitely generic structure for a theory T is a model of the finite forcing companion T^f.

Proof. Suppose M is a finitely generic structure for T. Let ϕ be a sentence in T^f. The sentence ϕ is defined in M. Suppose M does not satisfy ϕ. Then there is a condition P contained in the diagram of M such that $P \Vdash \sim\phi$. Since ϕ is in T^f, the empty condition weakly forces ϕ, i.e., $\emptyset \Vdash^* \phi$. But since $\emptyset \subseteq P$, $P \Vdash^* \phi$, contradicting that $P \Vdash \sim\phi$. Therefore, M satisfies ϕ.

A model M of a theory T is said to complete T if each model of T which extends M is an elementary extension of M. Equivalently, a model M of T completes T if and only if $T \cup \mathrm{Diag}(M)$ is a complete theory. Accordingly, a theory T is model-complete if and only if every model of T completes T. Barwise and Robinson (6) showed that the finitely generic structures are precisely those structures which complete the finite forcing companion.

Theorem 5.18. A structure is finitely generic relative to a theory T if and only if it is a model of T^f and completes T^f.

Proof. Suppose M is finitely generic relative to T. Then M is a model of T^f. Let M' extending M be a model of T^f. Let ϕ be a sentence defined in and satisfied by M. Then there is a condition P in the diagram of M such that P forces ϕ. If $a_0, \ldots, a_n$ are all the constants occurring in either ϕ or P but not in T, then T^f includes the sentence

$$\forall v_0 \ldots \forall v_n(\bigwedge P(v_0,\ldots, v_n) \to \phi(v_0,\ldots, v_n))$$

(see Proposition 5.10). Since P is in the diagram of M' and M' is a model of T^f, M' satisfies $\phi(a_0,\ldots, a_n)$. Thus, M' is an elementary extension of M.

Conversely, suppose M is a model of T^f and completes T^f. Proceed by induction on the complexity of sentences to show that M satisfies a sentence ϕ if and only if M forces ϕ. As usual, it suffices to consider sentences of the form $\sim\psi$. Suppose a sentence

$\sim\psi$ is defined in M. As in preceding proofs, if M forces $\sim\psi$, then M satisfies $\sim\psi$. Conversely, suppose M does not force $\sim\psi$. Then $T^f \cup \text{Diag}(M) \cup \{\psi\}$ is consistent, for let P be a finite subset of Diag(M). P is a condition, and since M does not force $\sim\psi$, there is a condition Q containing P such that Q forces ψ. If $a_0, \ldots, a_n$ are the constants occurring in either Q or ψ but not T, then T^f contains the sentence

$$\forall v_0 \ldots \forall v_n(\bigwedge Q(v_0,\ldots, v_n) \rightarrow \psi(v_0,\ldots, v_n)).$$

Therefore, $T^f \cup Q \cup \{\psi\}$ is consistent, so $T^f \cup P \cup \{\psi\}$ is consistent. Thus, $T^f \cup \text{Diag}(M) \cup \{\psi\}$ is consistent and has a model M'. Since M completes T^f and M' is an extension of M and a model of T^f and satisfies ψ, M satisfies ψ.

Theorem 5.19. If T is a theory in a countable language, then $T^f = \mathcal{T}\!h(\mathcal{F}_T)$.

Proof. According to Proposition 5.17, $\mathcal{T}\!h(\mathcal{F}_T)$ contains T^f. To prove the opposite inclusion let ϕ be a sentence in the language of T which is not in T^f. Then there is a countable, normal expansion $\mathcal{L}(A)$ and a condition P in $\mathcal{L}(A)$ such that P forces $\sim\phi$. Since $\mathcal{L}(A)$ is countable, there is a finitely generic structure M whose diagram contains P. Since P forces $\sim\phi$, M satisfies $\sim\phi$. Therefore, ϕ is not in $\mathcal{T}\!h(\mathcal{F}_T)$.

§ 4 Model-companions and Finite Forcing Companions

The finite forcing companion, like the infinite forcing companion, is a generalization of a model-companion. If a theory has a model-companion, then the infinite forcing companion and the finite forcing companion coincide with this model-companion. When no model-companion exists, the class of finitely generic models is the largest model-complete subclass which retains the constructive aspects of algebraically closed fields and whose theory is model-consistent with the original theory.

Proposition 5.20. Let T be a theory.

(i) T^f is model-complete if and only if every model of T^f is finitely generic relative to T.

(ii) T^f is model-complete if and only if every model of T^f is finitely generic relative to T^f.

(iii) T^f is a model-companion for T if and only if T^f is model-complete.

Proof. Part (i) is a direct consequence of theorem 5.18. Part (ii) follows from part (i) because $(T^f)_\forall = T_\forall$. Finally, since T^f and T are always mutually model-consistent, T^f is a model-companion if and only if it is model-complete.

Proposition 5.21. Let T be a theory. T^f is a companion theory for T, i.e.,

(1) $T^f = (T_\forall)^f$;

(2) $(T^f)_\forall = T_\forall$;

(3) $T^{ff} = T^f$;

(4) $(T_1)_\forall = (T_2)_\forall$ implies $T_1^f = T_2^f$;

(5) T^f is complete if and only if T has the joint embedding property;

(6) $T^f_{\forall\exists} = (\mathcal{T}h(\mathcal{E}_T))_{\forall\exists} \supseteq T_{\forall\exists}$;

(7) if T has a model-companion T^*, then T^f is the deductive closure of T^*.

Proof. Parts (1) and (2) have been established already. Parts (3) and (4) follow from (1) and (2). Part (5) follows from the fact that T has the joint embedding property if and only if whenever P

and Q are conditions relative to T such that the only constants occurring in both P and Q occur also in T, then $P \cup Q$ is a condition. Consequently, there cannot exist conditions P and Q and a sentence ϕ in the language of T such that $P \Vdash \phi$ and $Q \Vdash \sim\phi$. Therefore, for every sentence ϕ, either $\emptyset \Vdash \sim\phi$ or $\emptyset \Vdash \sim\sim\phi$.

To prove part (6), suppose that $\forall v_0 \ldots \forall v_n \; \exists v_{n+1} \ldots \exists v_m \; \psi(v_0, \ldots, v_n, v_{n+1}, \ldots, v_m)$ is a sentence in $\mathfrak{Th}(\mathcal{E}_T)$, where ψ is a quantifier-free formula. Suppose this sentence is not in T^f. Then there is a condition P which forces $\exists v_0 \ldots \exists v_n \; \forall v_{n+1} \ldots \forall v_m \sim\psi(v_0, \ldots, v_n, v_{n+1}, \ldots, v_m)$. So there are constants $a_0, \ldots, a_n$ such that P forces $\forall v_{n+1} \ldots \forall v_m \sim\psi(a_0, \ldots, a_n, v_{n+1}, \ldots, v_m)$. Let $P' = P \cup \{a_0 = a_0, \ldots, a_n = a_n\}$. P' is a condition, so P' is contained in the diagram of an existentially complete model M. By hypothesis, there are elements $b_1, \ldots, b_{m-n}$ such that M satisfies $\psi(a_0, \ldots, a_n, b_1, \ldots, b_{m-n})$. Let $Q = P' \cup \{\chi : \chi$ is a basic sentence in $\psi(a_0, \ldots, a_n, b_1, \ldots, b_{m-n})$ and χ is true in M$\}$. Then Q forces both $\exists v_{n+1} \ldots \exists v_m \; \psi(a_0, \ldots, a_n, v_{n+1}, \ldots, v_m)$ and $\forall v_{n+1} \ldots \forall v_m \sim\psi(a_0, \ldots, a_n, v_{n+1}, \ldots, v_m)$, which is impossible. Hence T^f contains $(\mathfrak{Th}(\mathcal{E}_T))_{\forall\exists}$.

To establish the reverse inclusion, suppose that ϕ is an $\forall\exists$ consequence of T^f. Let M be an existentially complete structure for T. Since $T^f_{\forall} = T_{\forall}$, there is an extension M' of M which is a model of T^f. Then ϕ holds in M' and M is existentially complete in M', so ϕ holds in M.

Finally, to prove (7), assume T has a model-companion T*. Then $T^*_{\forall\exists} = \mathfrak{Th}(\mathcal{E}_T)_{\forall\exists} = T^f_{\forall\exists}$. Since T* is axiomatized by $T^*_{\forall\exists}$, each theorem of T* is in T^f. Since T* is model-complete, every formula is equivalent in T* to an existential formula. The same equivalence must hold in T^f. Since T* and T^f have the same existential consequences, T* and T^f are logically equivalent.

Part (6) of the preceding proposition can be improved for countable theories.

Proposition 5.22. If T is a theory in a countable language, then T^f contains $\mathcal{TA}(\mathcal{E}_T)$.

Proof. When T is a theory in a countable language, then $T^f = \mathcal{TA}(\mathcal{F}_T)$ and $\mathcal{F}_T$ is a subclass of $\mathcal{E}_T$.

Proposition 5.23. A theory T is model-complete if and only if every model of T is finitely generic.

Proof. This follows from part (7) of Proposition 5.21, Theorem 5.18, and Proposition 5.14.

Thus, when a model-companion exists, the finite forcing companion is this model-companion and the finitely generic models are just the existentially complete structures. For example, for commutative fields, the finite forcing companion is the theory of algebraically closed fields, and the algebraically closed fields are precisely the finitely generic structures. Similar comments apply to real closed fields, Hensel fields, and algebraically closed abelian groups.

The notion of finite genericity captures in some sense the fact that the truth or falsity of every sentence defined in an algebraically closed field is determined by some finite amount of information about the parameters in the sentence. If T^f is a recursively enumerable theory and M is a finitely generic model whose diagram is recursively enumerable, then the truth or falsity of any sentence defined in M can be determined by simultaneously enumerating T^f and Diag(M) and comparing the finite subsets of Diag(M) with the sentences in T^f of the form in Proposition 5.10.

§ 5 Approximating Theories for the Finite Forcing Companion.

P. Henrard has discovered a construction for the finite forcing companion without the use of finite forcing. An ascending chain of theories is defined such that the nth theory in the chain is just the collection of all $\forall_n$ sentences in the finite forcing companion. Consequently, the finite forcing companion is the deductive closure of the union of this chain.

Some preliminary definitions and lemmas are required. An extension M' of a structure M is said to be an $\forall_n$ extension of M, denoted by $M \prec_{\forall_n} M'$, if each $\forall_n$ sentence defined and true in M is true in M' also.

Lemma 5.24. Let T and T' be two theories such that the language of T' includes the language of T. The set $T'_{\forall_{n+1}}$ of $\forall_{n+1}$ consequences of T' contains the set $T_{\forall_{n+1}}$ of $\forall_{n+1}$ consequences of T if and only if each model M' of T' has an $\forall_n$ extension which is a model of T.

Proof. Assume $T'_{\forall_{n+1}}$ contains $T_{\forall_{n+1}}$. Let M' be a model of T'. The desired structure M'' exists if and only if $\mathcal{T}h(M',\bar{m}')_{\forall_n} \cup T$ is consistent. Suppose that $\phi(a_0,\ldots, a_n)$ is a sentence in $\mathcal{T}h(M',\bar{m}')_{\forall_n}$ such that $T \cup \{\phi(a_0,\ldots, a_n)\}$ is inconsistent, where $a_0, \ldots, a_n$ are all of the constants in ϕ but not in T. Then T implies $\forall v_0 \ldots \forall v_n \sim\phi(v_0,\ldots, v_n)$, which is an $\forall_{n+1}$ sentence. By assumption, this sentence is in $T'_{\forall_{n+1}}$, contradicting that M' satisfies $\phi(a_0,\ldots, a_n)$.

Conversely, assume that each model M' of T' has an $\forall_n$ extension M'' which is a model T. Let $\forall v_0 \ldots \forall v_n\ \phi(v_0,\ldots, v_n)$ be an $\forall_{n+1}$ consequence of T, where ϕ is $\exists_n$. Let $a_0, \ldots, a_n$ be elements of M'. Then M'' satisfies $\phi(a_0,\ldots, a_n)$. Since M'' is an $\forall_n$ extension of M', M' satisfies $\phi(a_0,\ldots, a_n)$. Hence M

satisfies $\forall v_0 \ldots \forall v_n\ \phi(v_0, \ldots, v_n)$. Thus $T'_{\forall_{n+1}}$ includes $T_{\forall_{n+1}}$.

Corollary 5.25. An extension M' of M is an $\forall_{n+1}$ extension of M if and only if there is an $\forall_n$ extension M'' of M' which is an elementary extension of M.

Proof. Let $T = \mathcal{T\!h}(M,\bar{m})$, $T' = \mathcal{T\!h}(M',\bar{m}')$, and apply the preceding lemma.

Let T be a theory with language $\mathcal{L}$. Let T^0 be the set of quantifier-free consequences of T and $T^1 = T_\forall$. Define T^{n+1} for $n \geq 1$ by $T^{n+1} = \{\phi : \phi$ is an $\forall_{n+1}$ sentence in $\mathcal{L}$ and $T^n \cup \{\phi\}$ is mutually model-consistent with $T\}$. The following lemma shows that this definition is nonvacuous.

Lemma 5.26. For each $n < \omega$, T^{n+1} is mutually model-consistent with T and T^{n+1} contains T^n.

Proof. For $n = 0$, T^1 is trivially mutually model-consistent with T, and T^1 contains T^0. Let $n > 0$ and assume the lemma is true for all smaller ordinals. Then T^n is mutually model-consistent with T, so T^{n+1} contains T^n by definition. To show that T^{n+1} is mutually model-consistent with T, it suffices to show that if ϕ and ϕ' are $\forall_{n+1}$ sentences for which both $T^n \cup \{\phi\}$ and $T^n \cup \{\phi'\}$ are mutually model-consistent with T, then $T^n \cup \{\phi, \phi'\}$ is mutually model-consistent with T. Observe that $(T^n \cup \{\phi\})_{\forall_n} = T^n$, for if ψ is an $\forall_n$ consequence of $T^n \cup \{\phi\}$, then $T_\forall = (T^{n-1})_\forall \subseteq (T^{n-1} \cup \{\psi\})_\forall \subseteq (T^n \cup \{\phi\})_\forall = T_\forall$, i.e., $(T^{n-1} \cup \{\psi\})_\forall = T_\forall$, so ψ is in T^n. Similarly, $(T^n \cup \{\phi'\})_{\forall_n} = T^n$.

Let M_0 be a model of T. Since T and $T^n \cup \{\phi\}$ are mutually model-consistent, M_0 has an extension M_1 which is a model of $T^n \cup \{\phi\}$. Since $(T^n \cup \{\phi\})_{\forall_n} = T^n = (T^n \cup \{\phi'\})_{\forall_n}$, M_1 has an $\forall_{n-1}$ extension M_2 which is a model of $T^n \cup \{\phi'\}$. For the same reason, M_2 has an $\forall_{n-1}$ extension M_3 which is a model of $T^n \cup \{\phi\}$. Iterating this process yields a chain

$$M_0 \subseteq M_1 \prec_{\forall_{n-1}} M_2 \prec_{\forall_{n-1}} M_3 \prec_{\forall_{n-1}} M_4 \prec_{\forall_{n-1}} \cdots$$

such that M_{2i} is a model of $T^n \cup \{\phi'\}$ for $i > 0$ and M_{2i+1} is a model of $T^n \cup \{\phi\}$ for $i \geq 0$. Let $M = \bigcup_{i<\omega} M_i$. The structure M is an $\forall_{n-1}$ extension of M_i for each $i > 0$.

Let $\chi = \forall v_0 \ldots \forall v_r \exists v_{r+1} \ldots \exists v_m \psi(v_0, \ldots, v_r, v_{r+1}, \ldots, v_m)$ be an $\forall_{n+1}$ consequence of $T^n \cup \{\phi\}$, where ψ is an $\forall_{n-1}$ formula. Let $a_0, \ldots, a_r$ be elements of M. Choose j for which $a_0, \ldots, a_r \in M_{2j+1}$. Since M_{2j+1} satisfies χ, there are elements $a_{r+1}, \ldots, a_m$ of M_{2j+1} for which M_{2j+1} satisfies $\psi(a_0, \ldots, a_r, a_{r+1}, \ldots, a_m)$. Since M is an $\forall_{n-1}$ extension of M_{2j+1}, M satisfies $\psi(a_0, \ldots, a_r, a_{r+1}, \ldots, a_m)$. Hence M satisfies χ.

Similarly, M satisfies every $\forall_{n+1}$ consequence of $T^n \cup \{\phi'\}$. In particular, M is a model of $T^n \cup \{\phi, \phi'\}$. Thus, $T^n \cup \{\phi, \phi'\}$ is model-consistent with T. Since T^n contains $T_\forall$, $T^n \cup \{\phi, \phi'\}$ is mutually model-consistent with T.

Theorem 5.27 (Henrard (40)). For any theory T, $(T^f)_{\forall_n} = T^n$ for all $n < \omega$ and T^f is the deductive closure of $\bigcup_{n<\omega} T^n$.

Proof. The proof is by induction on n. Since $(T^f)_\forall = T_\forall$, the theorem is true for $n = 0, 1$. Assume $(T^f)_{\forall_n} = T^n$ for some $n \geq 1$. Let ϕ be on $\forall_{n+1}$ sentence. First, suppose ϕ is in T^f. The induction hypothesis implies that $T^n \cup \{\phi\}$ is contained T^f, so $T_\forall = (T^1)_\forall \subseteq (T^n)_\forall \subseteq (T^n \cup \{\phi\})_\forall \subseteq (T^f)_\forall = T_\forall$. Thus $T^n \cup \{\phi\}$ is mutually model-consistent with T, and ϕ is in T^{n+1}.

Conversely, suppose ϕ is not in T^f. The sentencc ϕ has the form $\forall v_0 \ldots \forall v_r \exists v_{r+1} \ldots \exists v_m \sim\psi(v_0, \ldots, v_r, v_{r+1}, \ldots, v_m)$, where ψ is an $\exists_{n-1}$ formula. Since ϕ is not in T^f, there is a condition P in a normal expansion $\mathcal{L}(A)$ of the language $\mathcal{L}$ of T such that

$$P \Vdash \exists v_0 \ldots \exists v_r \forall v_{r+1} \ldots \forall v_m \psi(v_0, \ldots, v_r, v_{r+1}, \ldots, v_m).$$

By definition, there are closed terms $t_0, \ldots, t_r$ in $\mathcal{L}(A)$ such that

$$P \Vdash \forall v_{r+1} \ldots \forall v_m \psi(t_0, \ldots, t_r, v_{r+1}, \ldots, v_m).$$

Let $a_1, \ldots, a_s$ be all of the constants in $\mathcal{L}(A)$ but not in $\mathcal{L}$ which occur in either P or $t_0, t_1, \ldots, t_n$. According to Proposition 5.10, T^f includes the sentence

$$\forall v_{m+1} \ldots \forall v_{m+s} (\bigwedge P(v_{m+1}, \ldots, v_{m+s}) \rightarrow \forall v_{r+1} \ldots \forall v_m \psi(t_0(v_{m+1}, \ldots, v_{m+s}), \ldots, t_r(v_{m+1}, \ldots, v_{m+s}), v_{r+1}, \ldots, v_m)).$$

This is an $\forall_n$ sentence, so by assumption it is in T^n. Since P is a condition relative to T, $T \cup P$ has a model M. Clearly, no extension of M can be a model of $T^n \cup \{\phi\}$. Hence, $T^n \cup \{\phi\}$ is not mutually model-consistent with T, so ϕ is not in T^n.

This theorem can be used to determine the finite forcing companion of theories of the form $T \cup \mathrm{Diag}(M)$ where M is existentially complete and to characterize theories which are their own finite forcing companions.

<u>Lemma 5.28</u>. Let M be a member of Σ_T. Then $(T_\forall \cup \mathrm{Diag}(M))_\forall = (\mathcal{T\!A}(M,\bar{m}))_\forall$ if and only if $(T_\forall \cup \mathrm{Diag}(M))^f = \mathcal{T\!A}(M,\bar{m})$.

<u>Proof</u>. The necessity of the first condition is clear. To prove sufficiency, assume that $(T_\forall \cup \mathrm{Diag}(M))_\forall = (\mathcal{T\!A}(M,\bar{m}))_\forall$. Let $T' = T_\forall \cup \mathrm{Diag}(M)$. The structure M is existentially complete for T' and is a substructure of every model of T'. Proceed by induction to show that $(T')^n = (\mathcal{T\!A}(M,\bar{m}))_{\forall_n}$ for all n. Since $(T')^1 = (T_\forall \cup \mathrm{Diag}(M))_\forall = (\mathcal{T\!A}(M,\bar{m}))_\forall$, equality holds for $n = 0,1$.

Assume that $(T')^n = (\mathcal{T}\!h(M,\bar{m}))_{\forall_n}$ for some $n \geq 1$. Let ϕ be an $\forall_{n+1}$ sentence in the language of M. Suppose ϕ is in $(\mathcal{T}\!h(M,\bar{m}))_{\forall_{n+1}}$ and let M' be a model of T'. Since M is an existentially complete substructure of M', there is an extension M'' of M' which is an elementary extension of M. Consequently, ϕ is true in M''. According to the induction hypothesis, M'' is also a model of $(T')^n$, so $(T')^n \cup \{\phi\}$ is model-consistent with T'. Since $(T')^n$ contains $(T')_\forall$, $(T')^n \cup \{\phi\}$ is mutually model-consistent with T'. Thus ϕ is in $(T')^{n+1}$. Conversely, suppose ϕ is in $(T')^{n+1}$. There is a model M' of $(T')^{n+1}$, which must extend M. Since $(\mathcal{T}\!h(M,\bar{m}))_{\forall_n} = (T')^n \subseteq (T')^{n+1}$, M' is an $\forall_n$ extension of M. Therefore M satisfies ϕ. Hence $(T')^{n+1} = (\mathcal{T}\!h(M,\bar{m}))_{\forall_{n+1}}$.

Corollary 5.29. If M is an existentially complete structure for T, then $(T \cup \mathrm{Diag}(M))^f = \mathcal{T}\!h(M,\bar{m})$.

Corollary 5.30. If M is existentially complete for T, then M is finitely generic for the theory $T \cup \mathrm{Diag}(M)$.

Corollary 5.31. If M is infinitely generic for T, then $(T \cup \mathrm{Diag}(M))^f = (T \cup \mathrm{Diag}(M))^F$.

A theory T is said to be forcing complete if T^f is the deductive closure of T. The following characterization of forcing complete theories is due to Barwise and Robinson (6); the proof, however, uses Henrard's more recent result.

Theorem 5.32. Suppose T is a theory in a language $\mathcal{L}$ such that whenever ϕ is a sentence in $\mathcal{L}$ and $T \cup \{\phi\}$ is consistent, then there is a model which completes T and satisfies ϕ. Then T is forcing complete. If T is a theory in a countable language, then this condition is also necessary.

<u>Proof</u>. One shows by induction that $T^n = T_{\forall_n}$ for all n. For $n = 0$ and $n = 1$, the equality follows directly from the definition of T^n. So assume that $T^n = T_{\forall_n}$ for some $n \geq 1$. Let ϕ be an $\forall_{n+1}$ sentence which is not $\forall_n$. To show that ϕ is in T^{n+1} if and only if ϕ is in $T_{\forall_{n+1}}$, it suffices to show that every model of T^n can be embedded in a model of $T^n \cup \{\phi\}$ if and only if $T \cup \{\sim\phi\}$ is inconsistent. First, suppose $T \cup \{\sim\phi\}$ is consistent. Then by hypothesis there is a model M which completes T and satisfies $\sim\phi$. By the induction hypothesis, T^n is just $T_{\forall_n}$, so M is a model of T^n. Suppose M were contained in a model M' of $T^n \cup \{\phi\}$. Since $(T^n \cup \{\phi\})_{\forall_n} \supseteq T^n = T_{\forall_n}$, there is a model M'' of T which is an $\forall_{n-1}$ extension of M'. Since M completes T,

$$\begin{array}{ccc} M & \prec & M'' \\ & \subseteq \quad & \succ_{\forall_{n-1}} \\ & M' & \end{array} .$$

Therefore, M' is an $\forall_n$ extension of M (Corollary 5.25), and M must satisfy ϕ (since ϕ is $\forall_{n+1}$ and $M \prec_{\forall_n} M'$). But this contradicts that M satisfies $\sim\phi$. Therefore, ϕ is not in T^{n+1}. Conversely, suppose $T \cup \{\sim\phi\}$ is inconsistent. Then $T^n \cup \{\phi\}$ is a set of consequences of T, so $(T^n)_\forall \subseteq (T^n \cup \{\phi\})_\forall \subseteq T_\forall = (T^n)_\forall$, and every model of T^n is a substructure of a model $T^n \cup \{\phi\}$. Since $T_\forall \subseteq T^n$, $T^n \cup \{\phi\}$ is mutually model-comsistent with T, and ϕ is in T^{n+1}.

For the second part of the theorem, assume T is a theory in a countable language. If $T \cup \{\phi\} = T^f \cup \{\phi\}$ is consistent, then the empty condition does not force $\sim\phi$, so some condition P forces ϕ. According to Theorem 5.11, there is a finitely generic model M which satisfies ϕ. By Theorem 5.18, M must complete $T^f = T$.

Corollary 5.33. Suppose T is a complete theory. If T has a model which completes T, then T is forcing complete. If T is a theory in a countable language, then this condition is necessary also.

CHAPTER 6

AXIOMATIZATIONS

If, for a first order theory T, any of the classes $\mathcal{E}_T$, $\mathcal{F}_T$, $\mathcal{G}_T$, or $\mathcal{A}_T$ is a generalized elementary class, then $\mathcal{E}_T = \mathcal{F}_T = \mathcal{G}_T$ and $\mathcal{M}(\mathcal{E}_T)$ is a model-companion for T. If the theory T does not have a model-companion, then these classes cannot be axiomatized in a first order language, but they can be axiomatized in an infinitary language. These infinitary axiomatizations are described in this chapter and are used to prove that every countable, $\aleph_0$-categorical theory has a model-companion, a result due to Saracino.

Proposition 6.1. Assume T is a theory in a countable language $\mathcal{L}$. Then

(i) (Simmons) $\mathcal{E}_T$ is axiomatizable in $\mathcal{L}_{\omega_1,\omega}$ by a countable set of sentences;

(ii) (Macintyre) $\mathcal{F}_T$ is axiomatizable in $\mathcal{L}_{\omega_1,\omega}$ by a countable set of sentences;

(iii) (Wood) $\mathcal{G}_T$ is axiomatizable in $\mathcal{L}_{\omega_1,\omega}$ by a set of sentences;

(iv) (Wood) $\mathcal{A}_T$ is axiomatizable in $\mathcal{L}_{\omega_1,\omega}$ by a set of sentences.

Proof. (i) For each universal formula $\phi(v_0,\ldots, v_n)$ in $\mathcal{L}$, let $\mathcal{S}_\phi = \{\psi(v_0,\ldots, v_n) : \psi(v_0,\ldots, v_n)$ is existential and $T_\forall \vdash \forall v_0 \ldots \forall v_n(\psi(v_0,\ldots, v_n) \rightarrow \phi(v_0,\ldots, v_n))\}$. Let Θ_ϕ be the sentence

$$\forall v_0 \ldots \forall v_n\ (\phi(v_0,\ldots, v_n) \leftrightarrow \bigvee_{\psi \in \mathcal{S}_\phi} \psi(v_0,\ldots, v_n)).$$

That the set of sentences $T_\forall \cup \{\Theta_\phi : \phi$ is a universal formula in $\mathcal{L}\}$ is a set of axioms for $\mathcal{E}_T$ follows from part (iii) of Proposition 1.6.

(ii) For each formula $\phi(v_0,\ldots, v_n)$ in $\mathcal{L}$, let

$\mathcal{P}_\phi = \{\psi(v_0,\ldots, v_n, v_{n+1},\ldots, v_m) : \psi$ is a finite conjunction of basic formulas in $\mathcal{L}$, and if $\mathcal{L}(A)$ is a normal expansion of A and P_ψ is the set of basic formulas in $\psi(a_0,\ldots, a_m)$, then P_ψ is a condition and $P_\psi \Vdash^* \phi(a_0,\ldots, a_n)\}$.

Let $\theta_\phi^!$ be the infinitary sentence

$$\forall v_0 \ldots \forall v_n(\phi(v_0,\ldots, v_n) \leftrightarrow \bigvee_{\psi\in\mathcal{P}_\phi} \exists v_{n+1} \ldots \exists v_m\, \psi(v_0,\ldots, v_n, v_{n+1},\ldots$$

That $T^f \cup \{\theta_\phi^! : \phi$ is a formula in $\mathcal{L}\}$ is a set of axioms for $\mathcal{F}_T$ follows from Proposition 5.10 and Theorem 5.18.

(iii) The infinitely generic structures are precisely the existentially complete structures for which satisfaction coincides with infinite forcing, or equivalently, those structures in which every finite set of elements realize a maximal existential type and in which satisfaction coincides with weak infinite forcing (Proposition 3.19). Since structures can be amalgamated over maximal existential types, for each formula $\phi(v_0,\ldots, v_n)$ and each maximal existential type $\Delta(v_0,\ldots, v_n)$, either Δ is in R_ϕ^* or Δ is in $R_{\sim\phi}^*$ but not both. For each maximal existential type $\Delta(v_0,\ldots, v_n)$ in R_ϕ^* let $\theta_{\phi,\Delta}$ be the sentence

$$\forall v_0 \ldots \forall v_n (\bigwedge\Delta(v_0,\ldots, v_n) \rightarrow \phi(v_0,\ldots, v_n)).$$

Then $T_\forall \cup \{\theta_\phi : \phi$ is a universal formula in $\mathcal{L}\} \cup \{\theta_{\phi,\Delta} : \phi$ is a formula in $\mathcal{L}$ and $\Delta \in R_\phi^*\}$ is a set of axioms in $\mathcal{L}_{\omega_1,\omega}$ for $\mathcal{G}_T$.

(iv) The existentially universal structures are precisely the existentially complete structures which also have the property that whenever $\Delta'(v_0,\ldots, v_n, v_{n+1},\ldots, v_m)$ is a maximal existential type extending the maximal existential type $\Delta(v_0,\ldots, v_n)$ and $a_0, \ldots, a_n$ realize $\Delta(v_0,\ldots, v_n)$, then there are elements $a_{n+1}, \ldots, a_m$ which realize $\Delta'(v_0,\ldots, v_n, v_{n+1},\ldots, v_m)$. For each pair of maximal existential types $\Delta(v_0,\ldots, v_n)$ and $\Delta'(v_0,\ldots, v_n, v_{n+1},\ldots, v_m)$ such that Δ' extends Δ, let $\theta_{\Delta,\Delta'}$ be the formula

$\forall v_0 \ldots \forall v_n(\bigwedge \Delta(v_0,\ldots, v_n) \rightarrow \exists v_{n+1} \ldots \exists v_m \bigwedge \Delta'(v_0,\ldots, v_n,v_{n+1},\ldots v_m))$

The set $T_\forall \cup \{\theta_\phi : \phi$ is a universal formula in $\mathcal{L}\}$

$\cup \{\theta_{\Delta,\Delta'} : \Delta'$ is a maximal existential type extending the maximal existential type $\Delta\}$

is a set of axioms for $\mathcal{A}_T$.

Corollary 6.2. If T is a theory in a countable, first order language $\mathcal{L}$, then each of the classes $\mathcal{E}_T$ and $\mathcal{F}_T$ is axiomatized by a sentence of $\mathcal{L}_{\omega_1,\omega}$.

Angus Macintyre (57) has shown that for groups and division rings, the sets of axioms in (iii) and (iv) have cardinality $2^{\aleph_0}$.

Part (iii) of the preceeding proposition refines a theorem of Robinson that the class of infinitely generic structures for a theory in a countable language can be axiomatized in $\mathcal{L}_{(2^{\aleph_0})^+,\omega}$ by a countable set of sentences. Namely, for each formula $\phi(v_0,\ldots, v_n)$ in $\mathcal{L}$, let Ψ_ϕ be the formula

$$\forall v_0 \ldots \forall v_n \ (\phi(v_0,\ldots, v_n) \leftrightarrow \bigvee_{\Delta \in R^*_\phi} \bigwedge \Delta(v_0,\ldots, v_n).$$

Ψ_ϕ asserts that whether or not $\phi(v_0,\ldots, v_n)$ is satisfied by elements $a_0, \ldots, a_n$ of a structure depends only on the existential type of $a_0, \ldots, a_n$. The set of sentences $T_\forall \cup \{\Psi_\phi : \phi$ is a formula in $\mathcal{L}\}$ is a set of axioms for $\mathcal{G}_T$. The converse of this observation also holds.

Theorem 6.3. (Robinson) If $\mathcal{C}$ is a subclass of Σ_T, is model-consistent with Σ_T, and is axiomatized within Σ_T by sentences Ψ'_ϕ where ϕ ranges over all formulas in $\mathcal{L}$, Ψ'_ϕ is the formula

$$\forall v_0 \ldots \forall v_n \ (\phi(v_0,\ldots, v_n) \leftrightarrow \bigvee_{\Delta \in \mathcal{S}'_\phi} \bigwedge \Delta(v_0,\ldots, v_n)),$$

and $\mathcal{S}'_\phi$ is a collection of existential $(n+1)$-types, then $\mathcal{C} = \mathcal{G}_T$.

Proof. It suffices to verify that the class $\mathcal{C}$ satisfies the conditions of Theorem 3.10. The first condition is part of the hypotheses on $\mathcal{C}$. Suppose, then, that M and M' are members of $\mathcal{C}$ with M' extending M, and suppose that M satisfies a sentence $\phi(a_0,\ldots, a_n)$. Since M is in $\mathcal{C}$, there is an existential type $\Delta(v_0,\ldots, v_n)$ in $\mathcal{S}'_\phi$ such that $a_0, \ldots, a_n$ realize Δ in M. Then $a_0, \ldots, a_n$ realize Δ in M', so since M' satisfies Ψ'_ϕ, M' satisfies $\phi(a_0,\ldots, a_n)$. Thus M is an elementary substructure of M.

Finally, suppose M' is in $\mathcal{C}$ and M is an elementary substructure of M. Let $a_0, \ldots, a_n$ be elements of M and let $\phi(v_0,\ldots, v_n)$ be a formula. Then M satisfies $\phi(a_0,\ldots, a_n)$ if and only if M' satisfies $\phi(a_0,\ldots, a_n)$, and M' satisfies $\phi(a_0,\ldots, a_n)$ if and only if M' satisfies $\bigvee_{\Delta\in\mathcal{S}'_\phi} \bigwedge\Delta(a_0,\ldots, a_n)$. Since each Δ is a collection of first order formulas and M is an elementary substructure of M', $a_0, \ldots, a_n$ realize Δ in M' if and only if they realize Δ in M. Thus, M' satisfies $\bigvee_{\Delta\in\mathcal{S}'_\phi} \bigwedge\Delta(a_0,\ldots, a_n)$ if and only if M satisfies $\bigvee_{\Delta\in\mathcal{S}'_\phi} \bigwedge\Delta(a_0,\ldots, a_n)$. Therefore M satisfies Ψ'_ϕ for all formulas ϕ in $\mathcal{L}$, so M is in $\mathcal{C}$.

If each of the formulas Ψ'_ϕ is finitary, then $\mathcal{U}_T$ is axiomatized in $\mathcal{L}$ and T has a model companion. This observation leads to the following theorem.

<u>Theorem 6.4</u>. If T is a theory in a countable first order language and if T has only finitely many maximal existential n-types for each n, then T has an $\aleph_0$-categorical model-companion.

<u>Proof</u>. Since T has only finitely many maximal existential n-types, for each maximal existential n-type $\Delta(v_0,\ldots, v_{n-1})$, one may choose an existential formula $\psi_\Delta(v_0,\ldots, v_{n-1})$ which is in Δ but is not any other maximal existential n-type $\Delta' \neq \Delta$. Also, for each formula $\phi(v_0,\ldots, v_n)$ in the language of T, the set R^*_ϕ contains

only a finite number of maximal existential types. For each formula ϕ, let Ψ''_ϕ be the formula

$$\forall v_0 \ldots \forall v_n(\phi(v_0,\ldots, v_n) \leftrightarrow \bigvee_{\Delta \in R^*_\phi} \psi_\Delta(v_0,\ldots, v_n)).$$

As each disjunction is finite and each formula $\psi_\Delta(v_0,\ldots, v_n)$ is first order, Ψ''_ϕ is an elementary sentence for each formula ϕ. Let $T^* = T_\forall \cup \{\Psi''_\phi : \phi$ is a formula in the language of $T\}$. Each infinitely generic model for T is a model of T^* (from the definition of R^*_ϕ and Proposition 3.19), so the class of models of T^* is a subclass of Σ_T and is model-consistent with Σ_T. Hence, by the preceding theorem, $\mathcal{M}(T^*) = \mathcal{Q}_T$ and T^* is a model-companion for T.

It remains to show that T^* is $\aleph_0$-categorical. First, note that any two mutually model-consistent theories have the same maximal existential types. Since T has only finitely many maximal existential n-types for each n, so does T^*. But for infinitely generic structures, the complete type of a finite sequence of elements is uniquely determined by their maximal existential type. Hence, T^* has only finitely many complete n-types for each n. By Ehrenfeucht's theorem, T^* is $\aleph_0$-categorical.

<u>Theorem 6.5</u> (Saracino). If T is a countable, $\aleph_0$-categorical theory without finite models, then T has an $\aleph_0$-categorical model-companion.

<u>Proof</u>. Since T is $\aleph_0$-categorical, it has only finitely many complete n-types for each n. Since every maximal existential n-type is contained in some complete n-type, and distinct maximal existential n-types are contained in distinct complete n-types, T has only finitely many maximal existential n-types for each n.

<u>Theorem 6.6</u>. If T is a countable theory, $T = T^F$, T has no finite models, and T is categorical in some infinite power, then T is model-complete.

Proof. If T is $\aleph_0$-categorical, then the result follows from the preceding theorem. Suppose, then that T is categorical in some uncountable power, hence in every uncountable power. Then every uncountable model of T is infinitely generic. Since every countable model is an elementary substructure of an uncountable model, every countable model is also infinitely generic. Hence every model of T is infinitely generic and T is model-complete.

Saracino (97) has shown that Theorem 6.5 does not hold if $\aleph_0$-categoricity is replaced by $\aleph_1$-categoricity and that Theorem 6.6 fails if $T = T^F$ is replaced by $T = T^f$ and categoricity is some infinite power is replaced by categoricity in some uncountable power.

M. Mortimer has proven the following results:

Theorem 6.7. If $\kappa \geq \omega$ and T is a κ-stable theory with a model-companion T*, then T* is κ-stable also.

Corollary 6.8. If T is stable (superstable) and has a model-companion T*, then T* is stable (superstable).

Theorem 6.9. If T is strongly minimal and has a model-companion T*, then T* is strongly minimal.

CHAPTER 7

FORCING AND RECURSION THEORY

Questions concerning recursiveness and forcing were investigated shortly after the introduction of finite and infinite forcing. The first application of finite forcing in algebra was in the proof by A. Macintyre that a finitely generated group is a subgroup of every existentially complete (i.e., algebraically closed) group if and only if the finitely generated group has a solvable word problem. The proof utilized finite forcing to omit the quantifier-free types of finitely generated groups with nonrecursive diagrams. A second question concerning recursion theory and forcing, raised by A. Robinson, asked for the degree of unsolvability of the infinite forcing companion of arithmetic. A. Robinson first showed that this degree was not recursively enumerable and later that this degree was not arithmetical. Subsequently, D. Goldrei, A. Macintyre, and H. Simmons proved that this degree was not analytical. Results in this chapter will show that this completely answers Robinson's question.

The initial goal of this chapter is the determination of upper bounds for the degrees of unsolvability of the theories T^f, T^F, and $\mathcal{T}h(\mathcal{E}_T)$ for a first order theory T. Then various assumptions--such as (i) T has the joint embedding property and $\mathcal{G}_T = \mathcal{E}_T$, (ii) T has the joint embedding property and has a countable, existentially universal model, or (iii) T has fewer than $2^{\aleph_0}$ many non-elementarily equivalent existentially complete structures--are shown to lower the upper bounds for the degrees of unsolvability of T^F and $\mathcal{T}h(\mathcal{E}_T)$.

When discussing second order arithmetic or recursion theory, we will use a second order language $\mathcal{L}^*$ with number variables $v_0, v_1, v_2, \ldots$, set variables $X_0, X_1, X_2, \ldots$, the binary number functions $+$ and $\cdot$, the binary number relation $<$, the binary relation $\in$ between numbers and sets, the constant symbols $0, 1, 2, \ldots,$ the connectives $\wedge, \vee, \sim$, the number quantifier $\exists$ and the set quantifier $\exists$. The universal quantifiers for number variables and set variables will be regarded as defined symbols, as will the bounded quantifiers $\exists x < y$ and $\forall x < y$. The symbols i, j, k, p, q, r, s and the symbols X, Y, Z will be used in place of subscripted number variables and set variables, respectively, so that formulas will be more readable The second order theory of arithmetic in the language $\mathcal{L}^*$ will be denoted by $\mathcal{TA}_2(\mathfrak{N})$.

A formula in $\mathcal{L}^*$ is said to be in prenex normal form if it consists of a sequence of quantifiers followed by a quantifier-free formula, all set quantifiers precede all number quantifiers, all unbounded number quantifiers precede all bounded number quantifiers, an there are no adjacent existential quantifiers and no adjacent universal quantifiers. Each formula in $\mathcal{L}^*$ is equivalent in the second order theory of arithmetic to a formula in prenex normal form.

For definitions of Σ^0_n, Π^0_n, Δ^0_n, Σ^s_n, Π^s_n, Δ^s_n, Σ^1_n, Π^1_n, and Δ^1_n formulas and sets and for definitions of other concepts of recursion theory, the reader should consult H. Rogers (87). Formulas which are Σ^0_n, Π^0_n, Δ^0_n, Σ^s_n, Π^s_n, or Δ^s_n will also be called arithmetical predic

In this chapter, T will denote a fixed theory in a countable language $\mathcal{L}$. Let $A = \{a_n : 1 \leq n < \omega\}$ be a set of new constants, and let $\mathcal{L}_r = \mathcal{L} \cup \{a_n : r|n\}$ for $r \neq 0$, $\mathcal{L}_0 = \mathcal{L}$ ($r|n$ means r divides n). Fix a recursive Gödel numbering of $\mathcal{L}_1$ such that $\ulcorner a_n \urcorner = 4n$, where $\ulcorner \sigma \urcorner$ denotes the Gödel number of the expression σ. The set of Gödel numbers of a set T' of formulas in $\mathcal{L}_1$ will be denoted by $\ulcorner T' \urcorner$. We assume for the sake of simplicity that the set $\ulcorner T \urcorner$ of Gödel numbers of sentences in T is an arithmetical set. If k is the Gödel number of a formula in $\mathcal{L}_1$, then ϕ_k will denote that formula. For definitions of the standard predicates

$\text{Formula}_{\mathcal{L}_r}(k)$, $\text{Sentence}_{\mathcal{L}_r}(k)$, Theorem(k,X) (i.e., X is the set of Godel numbers of a set of formulas which imply ϕ), etc., the reader should consult Shoenfield (103).

A theory in $\mathcal{L}_r$ is called a Henkin theory in $\mathcal{L}_r$ if for every formula $\phi(v_0)$ in $\mathcal{L}_r$ there is an integer n divisible by r such that the sentence $\exists v_0\ \phi(v_0) \to \phi(a_n)$ is deducible from the theory. A complete, Henkin theory T' in $\mathcal{L}_r$ uniquely determines a model $M_{T'}$ of T'. The elements of $M_{T'}$ are equivalence classes of constants of $\mathcal{L}_r$. The equivalence relation is defined by $b \backsim c$ for constants b and c if and only if $T' \vdash b = c$. Denote the equivalence class of b by $\bar{b}$. The equality $f(\bar{b}_0, \ldots, \bar{b}_{n-1}) = \bar{b}_n$ holds in $M_{T'}$ for an n-ary function symbol f in $\mathcal{L}_r$ if and only if $T' \vdash f(b_0, \ldots, b_{n-1}) = b_n$. The relation $R(\bar{b}_0, \ldots, \bar{b}_{n-1})$ holds in $M_{T'}$ for an n-ary relation symbol R in $\mathcal{L}_r$ if and only if $T' \vdash R(b_0, \ldots, b_n)$. The structure $M_{T'}$ with these functions and relations is a model of T' (see Shoenfield (103)).

§ 1 Degree of Unsolvability of T^F

An upper bound for the degree of unsolvability of T^F will be determined from upper bounds for the degrees of unsolvability of the theories $\mathcal{Th}(\mathcal{E}_n)$ and $\mathcal{Th}(\mathcal{H}_n)$, where $\mathcal{E}_n$ and $\mathcal{H}_n$ are the classes defined in Chapter 4.

Lemma 7.1. There is an arithmetical predicate $\mathcal{E}_0(X)$ such that $\mathcal{E}_0(S)$ holds for a set S of natural numbers if and only if S is the set of Godel numbers of a complete Henkin theory T' in one of the languages $\mathcal{L}_r$ and $(T')_\forall$ includes $T_\forall$.

Proof. Three bookkeeping predicates are required to keep track of the languages $\mathcal{L}_r$. Let L(k,q) be the formula

$\text{Formula}_{\mathcal{L}_1}(k) \wedge \forall i((4|i \wedge$ "$a_{i/4}$ occurs in ϕ_k") $\to 4q|i)$. The sentence $L(m,n)$ holds for integers m and n if and only if ϕ_m is a formula in $\mathcal{L}_n$. Let $\mathrm{Lang}(q,X)$ be the formula $\forall k((k \in X \to L(k,q)) \wedge \forall i > q \; \exists j \; (j \in X \wedge \sim L(j,i))$. The sentence $\mathrm{Lang}(n,S)$ holds if and only if S is the set of Godel numbers of a set of sentences in $\mathcal{L}_n$ and n is the largest integer for which this is true. Let $\mathrm{Def}(k,X)$ be the formula $\mathrm{Sentence}_{\mathcal{L}_1}(k) \wedge \exists i(\mathrm{Lang}(i,X) \wedge L(k,i))$. The sentence $\mathrm{Def}(n,S)$ holds if and only if there is an integer m such that $\mathrm{Lang}(m,S)$ holds and n is the Gödel number of a formula in $\mathcal{L}_m$.

Define predicates $P_1(X), \ldots, P_6(X)$ as follows:

$P_1(X)$ is $\forall p(p \in X \to \mathrm{Sentence}_{\mathcal{L}_1}(p))$;

$P_2(X)$ is $\forall p((\mathrm{Def}(p,X) \wedge \mathrm{Theorem}\,(p,X)) \to p \in X)$;

$P_3(X)$ is $\sim(\ulcorner \exists v_0 \; (v_0 \neq v_0)\urcorner \in X)$;

$P_4(X)$ is $\forall p(\mathrm{Def}(p,X) \to (p \in X \vee \ulcorner \sim\phi_p \urcorner \in X)$;

$P_5(X)$ is $\forall p \, \forall i((\mathrm{Def}(p,X) \wedge p = \ulcorner \exists v_0 \; \phi(v_0)\urcorner) \to \exists k \, \exists q \, (\mathrm{Lang}(q,X) \wedge 4q|k \wedge \ulcorner \exists v_0 \; \phi(v_0) \to \phi(a_{k/4})\urcorner \in X))$;

$P_6(X)$ is $\forall p((\mathrm{Sentence}_{\mathcal{L}_0}(p) \wedge \mathrm{Theorem}\,(p, \ulcorner T \urcorner) \wedge$ "ϕ_p is universal") $\to p \in X)$.

Let $\mathcal{E}_0(X)$ be the formula $P_1(X) \wedge P_2(X) \wedge P_3(X) \wedge P_4(X) \wedge P_5(X) \wedge P_6(X)$. The sentence $\mathcal{E}_0(S)$ holds for a set S of natural numbers if and only if S is the set of Godel numbers of a set of sentences in $\mathcal{L}_1$ $(P_1(S))$ which is closed under deduction $(P_2(S))$, is consistent $(P_3(S))$, is complete $(P_4(S))$, is Henkin $(P_5(S))$, and includes $T_\forall$ $(P_6(S))$. Each of the formulas $P_1(X)$, $P_2(X)$, $P_3(X)$, $P_4(X)$, and $P_5(X)$ is clearly arithmetical; and since the set $\ulcorner T \urcorner$ is arithmetical by assumption, $P_6(X)$ is an arithmetical formula. Consequently, $\mathcal{E}_0(X)$ is an arithmetical formula.

If $\mathcal{E}_0(S)$ holds, then S is the set of Godel numbers of a complete, Henkin theory T' in one of the languages $\mathcal{L}_r$. The model $M_{T'}$ will also be denoted by M_S. Note that, since $P_6(S)$ holds, M_S

is a structure in Σ_T.

Lemma 7.2. There is an arithmetical predicate $\mathcal{E}_1(X)$ such that

(i) if $\mathcal{E}_1(S)$ holds, then S is the set of Godel numbers of a complete, Henkin theory and M_S is an existentially complete structure for T; and

(ii) if M is a countable, existentially complete structure for T, then there is a set S of natural numbers such that $\mathcal{E}_1(S)$ holds and M is isomorphic to M_S.

Proof. Let Diag(p,X) be the formula $\mathcal{E}_0(X) \wedge \mathrm{Def}(p,X) \wedge$ "ϕ_p is a basic sentence" $\wedge\ p \in X$. The sentence Diag(n,S) holds if and only if $\mathcal{E}_0(S)$ holds and n is the Godel number of a basic sentence in the diagram of M_S. Let Consist(p,X) be the formula $\mathcal{E}_0(X) \wedge \mathrm{Def}(p,X) \wedge \sim\mathrm{Theorem}(\ulcorner \exists v_0(v_0 \neq v_0)\urcorner, \ulcorner T_\forall \urcorner \cup \{\phi_p\} \cup \{i : \mathrm{Diag}(i,X)\})$. The sentence Consist(n,S) holds if and only if $\mathcal{E}_0(S)$ holds and ϕ_n is a sentence defined in M_S and satisfied in some extension of M in Σ_T.

Let $P_7(X)$ be the formula $\forall p((\mathrm{Consist}(p,X) \wedge$ "ϕ_p is existential"$) \to p \in X)$. Let $\mathcal{E}_1(X)$ be the formula $\mathcal{E}_0(X) \wedge P_7(X)$.

Suppose $\mathcal{E}_1(S)$ holds. Then S determines a structure M_S in Σ_T with the property that any existential sentence defined in M_S and consistent with $T_\forall \cup \mathrm{Diag}(M_S)$ is true in M_S. According to Proposition 1.6, M_S is existentially complete.

Conversely, suppose M is a countable, existentially complete structure for T. Assign names from $\mathcal{L}_1$ to M so that each element of M is named by a constant in A. Let $T' = \mathcal{T}h_{\mathcal{L}_1}(M)$. If $S = \ulcorner T' \urcorner$, then $\mathcal{E}_1(S)$ holds, and M and M_S are isomorphic.

Next, predicates $\mathcal{E}_n(X)$ and $\mathcal{H}_n(X)$, which are related to the classes $\mathcal{E}_n$ and $\mathcal{H}_n$ in the same way as $\mathcal{E}_0(X)$ to $\mathcal{E}_T$, are defined. Let Ext(X,Y) be the formula

$\mathcal{E}_0(X) \wedge \mathcal{E}_0(Y) \wedge \exists k \exists q(\mathrm{Lang}(k,X) \wedge \mathrm{Lang}(q,Y) \wedge q \mid k \wedge q \neq k$
$\wedge \forall p((\mathrm{Def}(p,X) \wedge$ "ϕ_p is quantifier-free") $\rightarrow (p \in X \leftrightarrow p \in Y)))$.
Let $F_n'(p)$ be the recursive predicate which holds if and only if p is the Godel number of a substitution instance in $\mathcal{L}_1$ of a formula in the set F_n used to define $\mathcal{E}_n$ (see page 83). Assume that the formula $\mathcal{E}_n(X)$ for an $n \geq 1$ has been defined. Let $\mathcal{E}_{n+1}(X)$ be the formula

$\mathcal{E}_n(X) \wedge \forall i(\mathrm{Lang}(i,X) \rightarrow 2^n \mid i) \wedge \forall Y((\mathcal{E}_n(Y) \wedge \mathrm{Ext}(X,Y)) \rightarrow$
$\forall p((F_{n+1}(p) \wedge \mathrm{Def}(p,X) \rightarrow (p \in X \leftrightarrow p \in Y)))$.

Let $\mathcal{H}_0(X)$ and $\mathcal{H}_1(X)$ be the same formulas as $\mathcal{E}_0(X)$ and $\mathcal{E}_1(X)$, respectively. Assume that the formula $\mathcal{H}_n(X)$ for an $n \geq 1$ has been defined. Let $\mathcal{H}_{n+1}(X)$ be the formula

$\mathcal{H}_n(X) \wedge \forall i(\mathrm{Lang}(i,X) \rightarrow 2^{2n} \mid i) \wedge \forall Y(\mathcal{H}_n(Y) \wedge \mathrm{Ext}(X,Y)) \rightarrow \exists Z(\mathcal{H}_n(Z)$
$\wedge \mathrm{Ext}(Y,Z) \wedge \forall p(\mathrm{Def}(p,X) \rightarrow (p \in X \leftrightarrow p \in Z))))$.

<u>Lemma 7.3</u>. (i) For $n \geq 1$, $\mathcal{E}_n(X)$ is a Π^1_{n-1} formula and (a) if $\mathcal{E}_n(S)$ holds, then S is the set of Godel numbers of a complete, Henkin theory and M_S is in $\mathcal{E}_n$, and (b) if M is a countable structure in $\mathcal{E}_n$, then there is a complete, Henkin theory T' in $\mathcal{L}_{2n}$ such that $\mathcal{E}_n(\ulcorner T' \urcorner)$ holds and M is isomorphic to $M_{T'}$.

(ii) For $n \geq 1$, $\mathcal{H}_n(X)$ is a Π^1_{2n-2} formula and (a) if $\mathcal{H}_n(S)$ holds, then S is the set of Godel numbers of a complete, Henkin theory and M_S is in $\mathcal{H}_n$, and (b) if M is a countable structure in $\mathcal{H}_n$, then there is a complete, Henkin theory T' in $\mathcal{L}_{2^{2n}}$ such that $\mathcal{H}_n(\ulcorner T' \urcorner)$ holds and M is isomorphic to $M_{T'}$.

<u>Proof</u>. Assume that part (i) is true for some $n \geq 1$. By inspection, $\mathcal{E}_{n+1}$ is a Π^1_n formula. Suppose $\mathcal{E}_{n+1}(S)$ holds. Then S is the set of Godel numbers of a complete, Henkin theory in some language $\mathcal{L}_r$ where $2^{n+1} \mid r$, and M_S is a member of $\mathcal{E}_n$. Suppose M' is an extension of M_S in $\mathcal{E}_n$. Replacing M' by an elementary substructure if necessary, we may assume that M' is countable.

Assign names from $\mathcal{L}_{r/2}$ to M' so that each member of M' is named by a constant in $\{a_i : (r/2)|i\}$ and the assignment extends the assignment of names in $\mathcal{L}_r$ to M. Let T' be the complete theory of M' in the language $\mathcal{L}_{r/2}$. According to the induction hypothesis, $\mathcal{E}_n(\ulcorner T' \urcorner)$ holds, and $\mathrm{Ext}(S, \ulcorner T' \urcorner)$ holds also. Since $\mathcal{E}_{n+1}(S)$ holds, each substitution instance ϕ_p of a formula in F_{n+1} holds in M_S if and only if it holds in M'. Hence, M_S is F_{n+1}-persistently complete and is in $\mathcal{E}_{n+1}$.

Conversely, suppose M is a countable member of $\mathcal{E}_{n+1}$. Assign names from $\mathcal{L}_{2n+1}$ to M such that each element of M is named by a constant in the set $\{a_i : 2^{n+1}|i\}$. Let T' be the complete theory of M in $\mathcal{L}_{2(n+1)}$. If S is a set for which $\mathcal{E}_n(S)$ holds and $\mathrm{Ext}(\ulcorner T' \urcorner, S)$ holds, then S determines a model M_S in $\mathcal{E}_n$ such that M_S extends M. Since M is F_{n+1}-persistently complete, $\mathcal{E}_{n+1}(\ulcorner T' \urcorner)$ holds, and M and $M_{T'}$ are isomorphic.

The proof of part (ii) is similar. The technical requirement that $2^{2n}|r$ is necessary because the definition of $\mathcal{H}_{n+1}$ from $\mathcal{H}_n$ refers to two successive extensions of a member of $\mathcal{H}_n$.

<u>Lemma 7.4</u>. For $n \geq 1$, (i) $\ulcorner \mathcal{TA}(\mathcal{E}_n) \urcorner$ is a Π^1_n set,
(ii) $\{p : \sim\phi_p \in \mathcal{TA}(\mathcal{E}_n)\}$ is a Π^1_n set, and
(iii) $\{p : \phi_p \notin \mathcal{TA}(\mathcal{E}_n)$ and $\sim\phi_p \notin \mathcal{TA}(\mathcal{E}_n)\}$ is a Σ^1_n set.

<u>Proof.</u> According to preceding lemmas, the formula $\mathcal{E}_n(X)$ is arithmetical for $n = 1$ and is Π^1_{n-1} for $n > 1$. The set of Godel numbers in (i) is just $\{p : \mathrm{Sentence}_{\mathcal{L}}(p) \wedge \forall X(\mathcal{E}_n(X) \rightarrow p \in X)\}$, which is a Π^1_n set. Similarly, the set in part (ii) is just $\{p : \mathrm{Sentence}_{\mathcal{L}}(p) \wedge \forall X(\mathcal{E}_n(X) \rightarrow \ulcorner \sim\phi_p \urcorner \in X\}$, which is a Π^1_n set also. Finally, the set in part (iii) is the set
$\{p : \mathrm{Sentence}_{\mathcal{L}}(p) \wedge \exists X \exists Y(\mathcal{E}_n(X) \wedge \mathcal{E}_n(Y) \wedge \sim(p \in X) \wedge \sim(\ulcorner \sim\phi_p \urcorner \in Y))\}$, which is a Σ^1_n set.

Lemma 7.5. For $n \geq 1$, (i) $\ulcorner \mathcal{TA}(\mathcal{H}_n) \urcorner$ is a Π^1_{2n-1} set,
(ii) $\{p : \sim\phi_p \in \mathcal{TA}(\mathcal{H}_n)\}$ is a Π^1_{2n-1} set, and
(iii) $\{p : \phi_p \notin \mathcal{TA}(\mathcal{H}_n)$ and $\sim\phi_p \notin \mathcal{TA}(\mathcal{H}_n)\}$ is a Σ^1_{2n-1} set.

Proof. Analogous to the preceding proof.

Lemmas 7.4 and 7.5 yield the following bound on the degree of unsolvability of T^F.

Theorem 7.6. T^F is one-one reducible to $\mathcal{TA}_2(\mathcal{N})$.

Proof. Let ϕ be a sentence in $\mathcal{L}$. The sentence ϕ is logically equivalent to a sentence ϕ' of $\mathcal{L}$ in prenex normal form. Choose an $n > 0$ such that ϕ' is either an $\exists_n$ or an $\forall_n$ formula. Then ϕ' is persistent under both extension and restriction in the class $\mathcal{E}_{n+1}$. Since $\mathcal{E}_{n+1}$ includes the class $\mathcal{G}_T$ and $\mathcal{G}_T$ is model-consistent with $\mathcal{E}_{n+1}$, ϕ' is in $T^F = \mathcal{TA}(\mathcal{G}_T)$ if and only if ϕ' is in $\mathcal{TA}(\mathcal{E}_{n+1})$. Thus, $\ulcorner\phi\urcorner$ is in $\ulcorner T^F \urcorner$ if and only if $\ulcorner\phi'\urcorner$ is in $\ulcorner \mathcal{TA}(\mathcal{E}_{n+1})\urcorner$, that is, $\ulcorner\phi\urcorner$ is in $\ulcorner T^F\urcorner$ if and only if the sentence "$\ulcorner\phi'\urcorner \in \ulcorner\mathcal{TA}(\mathcal{E}_{n+1})\urcorner$" is a theorem in $\mathcal{TA}_2(\mathcal{N})$. Clearly, this reduction is effective, so T^F is one-one reducible to $\mathcal{TA}_2(\mathcal{N})$.

§ 2 Degree of Unsolvability of T^f

The theories T^n defined in Chapter 5 will be used for the determination of an upper bound for the degree of unsolvability of T^f. Define inductively a sequence of predicates $\mathcal{T}^n(p)$ according to the following schema:

$\mathcal{T}^1(p)$ is $\text{Sentence}_{\mathcal{L}}(p) \wedge$ "ϕ_p is an $\forall_1$ formula"
Theorem (p, T),

$\mathcal{T}^{n+1}(p)$ is $\text{Sentence}_{\mathcal{L}}(p) \wedge$ "ϕ_p is an $\forall_{n+1}$ formula"
$\wedge \sim\text{Theorem}\,(\ulcorner\exists v_0(v_0 \neq v_0)\urcorner, \{p\} \cup \{q : \mathcal{T}^n(q) \text{ holds}\})$

$\wedge \quad \forall k((\text{Sentence}_{\mathcal{L}}(k) \wedge \text{"}\phi_k \text{ is an } \forall_1 \text{ formula"})$

$\rightarrow (\text{Theorem } (k, \ulcorner T \urcorner) \leftrightarrow \text{Theorem}(k, \{p\} \cup \{q : \mathcal{T}^n(q) \text{ holds}\}))).$

Since $\ulcorner T \urcorner$ is an arithmetical set, $\mathcal{T}^1(p)$ is an arithmetical formula and by induction so is $\mathcal{T}^n(p)$ for all $n \geq 1$. Moreover, an $\forall_n$ sentence ϕ_p is in T^n if and only if $\mathcal{T}^n(p)$ is true.

Theorem 7.7. T^f is one-one reducible to $\mathcal{TA}(N)$.

Proof. Let ϕ be a sentence in $\mathcal{L}$. The sentence ϕ is logically equivalent to a sentence ϕ' in prenex normal form. Choose the least $n \geq 1$ for which ϕ' is either an $\exists_n$ or an $\forall_n$ formula. Since T^f is the deductive closure of $\bigcup_{n<\omega} T^n$, ϕ is in T^f if and only if ϕ' is in T^{n+1}. In other words, $\ulcorner \phi \urcorner$ is in $\ulcorner T^f \urcorner$ if and only if $\mathcal{T}^{n+1}(\ulcorner \phi' \urcorner)$ is a true sentence in N. Clearly this is an effective reduction, so T^f is one-one reducible to $\mathcal{TA}(N)$.

Corollary 7.8. If T^F is not a hyperarithmetical set, then the class of infinitely generic structures and the class of finitely generic structures do not coincide, and T has no model-companion.

Results in Part Two on models of arithmetic and in Part Three on division rings show that the bounds in Lemma 7.4, Lemma 7.5, Theorem 7.6, and Theorem 7.7 are the best possible.

§ 3 Consequences of the Joint Embedding Property

If T has the joint embedding property, then any two infinitely generic structures are elementarily equivalent and T^F is a complete theory. This together with other hypotheses can reduce the upper bound on the degree of unsolvability of T^F.

Proposition 7.9. Assume T has the joint embedding property. If $T^F = \mathcal{TA}(\mathcal{E}_n)$, then $\ulcorner T^F \urcorner$ is a Δ^1_n set. If $T^F = \mathcal{TA}(\mathcal{H}_n)$, then

$\ulcorner T^F \urcorner$ is a Δ^1_{2n-1} set.

Proof. Suppose $T^F = \mathcal{TA}(\mathcal{E}_n)$. Since T has the joint embedding property and $T^F = \mathcal{TA}(\mathcal{E}_n)$, $\mathcal{TA}(\mathcal{E}_n)$ is a complete theory. Consequent the set $\{p : \phi_p \notin \mathcal{TA}(\mathcal{E}_n)$ and $\sim\phi_p \notin \mathcal{TA}(\mathcal{E}_n)\}$ is empty. Then the complement of $\{p : \phi_p \in \mathcal{TA}(\mathcal{E}_n)\}$ is just $\{p : \sim\phi_p \in \mathcal{TA}(\mathcal{E}_n)\} \cup \{p :$ p is not the Godel number of a sentence of $\mathcal{L}\}$, which is a Π^1_n set. Therefore, $\ulcorner \mathcal{TA}(\mathcal{E}_n) \urcorner$ is both Π^1_n and Σ^1_n, so $\ulcorner \mathcal{TA}(\mathcal{E}_n) \urcorner = \ulcorner T^F \urcorner$ is a Δ^1_n set.

A similar argument holds when $T^F = \mathcal{TA}(\mathcal{H}_n)$.

Corollary 7.10. Assume T has the joint embedding property. If $T^F = \mathcal{TA}(\mathcal{E}_T)$, then $\ulcorner T^F \urcorner$ is hyperarithmetical.

Corollary 7.11. Assume T has the joint embedding property. If $\mathcal{G}_T = \mathcal{E}_T$, then $\ulcorner T^F \urcorner = \ulcorner \mathcal{TA}(\mathcal{G}_T) \urcorner = \ulcorner \mathcal{TA}(\mathcal{E}_T) \urcorner$ is hyperarithmetical.

Corollary 7.12. If R is a countable ring whose diagram is an arithmetical set, then the theory of the existentially complete R-modules is hyperarithmetical.

When T has the joint embedding property, the degree of unsolvabili of T^F is a lower bound for the degree of unsolvability of the diagrams of countable, infinitely generic structures.

Proposition 7.13. Assume T has the joint embedding property. If T has a countable, infinitely generic structure whose diagram is a Σ^1_n, Π^1_n or Δ^1_n set $(n > 0)$, then $\ulcorner T^F \urcorner$ is Σ^1_n, Π^1_n or Δ^1_n, respectively. In particular, if there is a countable, infinitely generic structure whose diagram is arithmetical, then $\ulcorner T^F \urcorner$ is hyperarithmetical.

Proof. Suppose T has a countable, infinitely generic structure M whose diagram is a Σ^1_n or a Π^1_n set, i.e., the diagram of M under some assignment of names from $\mathcal{L}_1$ to M is a Σ^1_n or a Π^1_n

set. Let $D(p)$ be the Σ_n^1 or Π_n^1 formula such that $D(m)$ holds if and only if m is the Gödel number of a basic sentence in the diagram of M. If $D(p)$ is a Σ_n^1 formula, then ϕ_p is in T^F if and only if

$$\text{Sentence}_{\mathcal{L}}(p) \wedge \exists X(\mathcal{E}_0(X) \wedge \text{Lang}(1,X) \wedge \forall i(\text{Basic Sentence}_{\mathcal{L}_1}(i) \rightarrow ((i \in X) \rightarrow D(i))) \wedge p \in X).$$

Therefore, T^F is a Σ_n^1 set.

If $D(p)$ is a Π_n^1 formula, then ϕ_p is in T^F if and only if

$$\text{Sentence}_{\mathcal{L}}(p) \wedge \forall X((\mathcal{E}_0(X) \wedge \text{Lang}(1,X) \wedge \forall i(\text{Basic Sentence}_{\mathcal{L}_1}(i) \rightarrow (D(i) \rightarrow i \in X))) \rightarrow p \in X).$$

Therefore, T^F is a Π_n^1 set.

If T has a countable, infinitely generic structure whose diagram is Δ_n^1, then this diagram is both Σ_n^1 and Π_n^1, so T^F is both Σ_n^1 and Π_n^1.

The existence of a countable, existentially universal structure for a theory with the joint embedding property lowers the upper bound on the degree of unsolvability of the infinite forcing companion also.

<u>Lemma 7.14</u>. There is a Π_1^1 formula $\mathcal{EU}(X)$ such that

(i) if $\mathcal{EU}(S)$ holds, then S is the set of Gödel numbers of a complete, Henkin theory in $\mathcal{L}_2$ and M_S is existentially universal for T, and

(ii) if M is countable and existentially universal, then there is a complete, Henkin theory T' in $\mathcal{L}_2$ such that $\mathcal{EU}(\ulcorner T' \urcorner)$ holds and M and $M_{T'}$ are isomorphic.

<u>Proof</u>. Let $\text{Subst}(p, r, k, q)$ be the predicate which asserts that if ϕ_p is a formula of $\mathcal{L}_1$ and k is the Gödel number of a constant b of $\mathcal{L}_1$, then r is the Gödel number of the formula obtained by replacing each occurrence of b in ϕ_p by an occurrence of a_q. Let $\text{Existtype}(p, i, j)$ be the formula

$\text{Constant}_{\mathcal{L}_1}(i) \wedge \text{Sentence}_{\mathcal{L}_1}(p) \wedge$ "ϕ_p is an existential formula"

$\wedge\ \forall s(\text{Constant}_{\mathcal{L}_1}(s) \to ((\text{"the constant } b \text{ with } \ulcorner b \urcorner = s$

occurs in ϕ_p") $\to (s = i \vee (8|s \wedge s \leq j))))$.

Existtype(p, i, j) holds if and only if ϕ_p is an existential sentence and the only constants possibly occurring in ϕ_p are the constant b for which $\ulcorner b \urcorner = i$ and constants from $A \cap \mathcal{L}_2$ with subscript $\leq j/4$.

Let $\mathcal{EU}(X)$ be the formula

$\mathcal{E}_1(X) \wedge \text{Lang}(2,X) \wedge \forall Y((\mathcal{E}_1(Y) \wedge \text{Ext}(X,Y)) \to \forall i \forall j \exists k\ (8|k$

$\wedge\ \forall p((\text{Existtype}(p, i, j) \wedge p \in Y) \to \forall r(\text{Subst}(p, r, i, k)$

$\to r \in X))))$.

The formula $\mathcal{EU}(X)$ is a Π^1_1 formula.

Suppose $\mathcal{EU}(S)$ holds for a set S of natural numbers. Then S is the set of Godel numbers of a complete, Henkin theory in $\mathcal{L}_2$, and M_S is existentially complete. Suppose M' is a countable, existentiall complete extension of M_S. Extend the assignment of names from $\mathcal{L}_2$ to M_S to an assignment of names from $\mathcal{L}_1$ to M'. Since $\mathcal{EU}(S)$ holds, each existential one-type defined in M_S and realized in M' is realized in M_S. Therefore, M_S is existentially universal for T (Proposition 1.20).

Conversely, suppose M is a countable, existentially universal structure for T. Assign names from $\mathcal{L}_2$ to M such that each element of M is named by a constant in A. Let T' be the complete theory of M in the language $\mathcal{L}_2$ relative to this assignment. Then $\mathcal{EU}(\ulcorner T' \urcorner)$ holds and M and $M_{T'}$ are isomorphic.

<u>Proposition 7.15</u>. If T has the joint embedding property and T has a countable, existentially universal structure, then $\ulcorner T^F \urcorner$ is a Δ^1_2 set.

<u>Proof</u>. Since T has the joint embedding property, all existentiall universal structures are elementarily equivalent and T^F is the theory of the existentially universal structures. Since T has a countable,

existentially universal structure, the following are equivalent: ϕ_p is a sentence in T^F; $\forall X(\mathcal{EU}(X) \to p \in X)$; $\exists X(\mathcal{EU}(X) \wedge p \in X)$. Consequently, $\ulcorner T^F \urcorner$ is a Δ^1_2 set.

§ 4 Non-elementarily Equivalent Existentially Complete Structures

Theorem 7.17 was suggested by A. Macintyre's result (58) that there are $2^{\aleph_0}$ non-elementarily equivalent, existentially complete division rings in each characteristic different from two. A. Macintyre's construction of these division rings utilized finite forcing and an infinite collection of arithmetical sets, no one of which was Turing reducible to a finite collection of the others. In general, such a direct construction is no longer necessary. Indeed, as Theorem 7.17 asserts, there exist $2^{\aleph_0}$ non-elementarily equivalent, existentially complete structures if and only if there exists one existentially complete structure whose theory is not a hyperarithmetical set.

The crux of the proof is the observation, due to J. Hirschfeld, that a complete theory T' is the theory of an existentially complete structure if and only if it includes the universal consequences of T and certain types are nonprincipal in T'.

Lemma 7.16. A complete, consistent theory T' with the same language as T is the theory of an existentially complete structure for T if and only if

(i) $(T')_\forall$ includes $T_\forall$, and

(ii) for each universal formula $\phi(v_0,\ldots, v_n)$ in the language of T, the following type is nonprincipal in T':

$\{\phi(v_0,\ldots, v_n)\} \cup \{\sim\psi(v_0,\ldots, v_n) : \psi$ is existential and $T_\forall \vdash \forall v_0 \ldots \forall v_n(\psi(v_0,\ldots, v_n) \to \phi(v_0,\ldots, v_n))\}$.

Proof. Let $\mathcal{U}$ denote the collection of types described in the lemma. According to Proposition 1.6, a structure is existentially complete relative to T if and only if it is a model of $T_\forall$ and omits each of the types in $\mathcal{U}$. Consequently, if T' is the theory of an existentially complete structure M, then T' includes $T_\forall$ and each of the types in $\mathcal{U}$ is nonprincipal in T (because M omits each of them).

Conversely, suppose T' is a complete theory satisfying conditions (i) and (ii). Then T' has a countable model M omitting each of the types in $\mathcal{U}$. Since T' includes $T_\forall$, M is in Σ_T. Since M omits each type in $\mathcal{U}$, M is existentially complete.

Theorem 7.17. If the set $\mathcal{R} = \{\ulcorner T' \urcorner : T' = \mathcal{T}h(M)$ for some existentially complete structure $M\}$ has cardinality less than $2^{\aleph_0}$, then $\mathcal{R}$ is countable, each member of $\mathcal{R}$ is hyperarithmetical, and $\ulcorner \mathcal{T}h(\mathcal{E}_T)\urcorner$ is hyperarithmetical.

Proof. Let $\mathcal{R}(X)$ be the formula

$\forall p(p \in X \rightarrow \text{Sentence}_{\mathcal{L}}(p)) \wedge \forall p((\text{Sentence}_{\mathcal{L}}(p) \wedge \text{Theorem}(p,X)) \rightarrow p \in$
$\wedge\ \forall p(\text{Sentence}_{\mathcal{L}}(p) \rightarrow (p \in X \vee \ulcorner \sim\phi_p \urcorner \in X) \wedge P_3(X) \wedge P_6(X)$
$\wedge\ \forall p(($"ϕ_p is a universal formula of $\mathcal{L}$ with free variables among $v_0, \ldots, v_n$") $\rightarrow \forall k(($"ϕ_k is a formula of $\mathcal{L}$ with free variables among $v_0, \ldots, v_n$") $\rightarrow (\sim(\ulcorner \exists v_0 \ldots \exists v_n\, \phi_k(v_0, \ldots, v_n)\urcorner \in$
$\vee \sim(\ulcorner \forall v_0 \ldots \forall v_n(\phi_k(v_0, \ldots, v_n) \rightarrow \phi_p(v_0, \ldots, v_n)\urcorner \in X)$
$\vee \sim(\forall q((($"ϕ_q is an existential formula of $\mathcal{L}$ with free variables among $v_0, \ldots, v_n$") $\wedge$ Theorem$(\ulcorner \forall v_0 \ldots \forall v_n(\phi_q(v_0, \ldots, v$
$\rightarrow \phi_p(v_0, \ldots, v_n)\urcorner, \ulcorner T_\forall \urcorner)) \rightarrow (\ulcorner \forall v_0 \ldots \forall v_n(\phi_k(v_0, \ldots, v_n)$
$\rightarrow \sim\phi_q(v_0, \ldots, v_n)\urcorner \in X)))))$.

$\mathcal{R}(S)$ holds for a set S of natural numbers if and only if S is the set of Godel numbers of a complete theory T' which includes $T_\forall$ and in which each of the types described in the preceding lemma is nonprincipal. In other words, $\mathcal{R}(S)$ holds if and only if S is in $\mathcal{R}$

The formula $\mathcal{R}(X)$ has no set quantifiers, so the collection of sets of natural numbers satisfying $\mathcal{R}(X)$ is an arithmetical collection. A standard result of recursion theory (see Sacks (92)) asserts that if a hyperarithmetical collection of sets has fewer than $2^{\aleph_0}$ members, then the collection is countable, each member is hyperarithmetical, and the collection is indexed by a hyperarithmetical function. This proves the first and second parts of the theorem.

To verify the last part, one observes that $\ulcorner \mathcal{TA}(\mathcal{E}_T) \urcorner = \bigcap \{S : \mathcal{R}(S)$ holds$\}$. Since the intersection of a collection of hyperarithmetical sets indexed by a hyperarithmetical function is hyperarithmetical, $\ulcorner \mathcal{TA}(\mathcal{E}_T) \urcorner$ is hyperarithmetical if $\mathcal{R}$ has fewer than $2^{\aleph_0}$ members.

S. Simpson suggested the use of the result on collections of hyperarithmetical sets instead of direct construction of $2^{\aleph_0}$ distinct theories.

<u>Corollary 7.18</u>. If the set $\mathcal{R}$ has cardinality less than $2^{\aleph_0}$, then the theory of each infinitely generic structure is hyperarithmetical.

<u>Corollary 7.19</u>. If T has the joint embedding property, then either T^F is a hyperarithmetical set or T has $2^{\aleph_0}$ non-elementarily equivalent, existentially complete structures (or both).

SUMMARY

Three analogues for the class of algebraically closed fields -- the class $\mathcal{E}_T$ of existentially complete structures, the class $\mathcal{F}_T$ of finitely generic structures, and the class $\mathcal{G}_T$ of infinitely generic structures -- and one analogue for the class of universal domains -- the class $\mathcal{A}_T$ of existentially universal structures -- have been defined for a first order theory T. These classes have been analyzed in terms of extensiveness, the persistence of elementary properties under extension and restriction, and closure under substructures. The basic results are

(1) each of the classes $\mathcal{E}_T$, $\mathcal{G}_T$, and $\mathcal{A}_T$ are model-consistent with Σ_T;

(2) each of the classes $\mathcal{F}_T$, $\mathcal{G}_T$, and $\mathcal{A}_T$ are model-complete;

(3) each of the classes $\mathcal{F}_T$ and $\mathcal{E}_T$ are closed under existentially complete substructures;

(4) the class $\mathcal{G}_T$ is closed under elementary substructures;

(5) the class $\mathcal{A}_T$ is closed under $\mathcal{L}_{(\text{card } T)^+,\omega}$ - substructures;

(6) $\Sigma_T \supseteqq \mathcal{E}_T \supseteqq \mathcal{G}_T \supseteqq \mathcal{A}_T$; $\mathcal{E}_T \supseteqq \mathcal{F}_T$

(7) the class $\mathcal{E}_T$ is the unique subclass of Σ_T which is model-consistent with Σ_T, in which inclusion of one structure in a second entails that the first is existentially complete in the second, and which is closed under existentially complete substructures;

(8) the class $\mathcal{G}_T$ is the unique subclass of Σ_T which is model-consistent with Σ_T, is model-complete, and is closed under elementary substructures.

The theories of the classes $\mathcal{E}_T$, $\mathcal{F}_T$, $\mathcal{G}_T$, and $\mathcal{A}_T$ are analogues of a model-companion for the theory T. The basic results are

(1) T has a model-companion T* if and only if one of the classes $\mathcal{E}_T$, $\mathcal{F}_T$, $\mathcal{G}_T$, or $\mathcal{A}_T$ is axiomatizable, in which case $\mathcal{E}_T = \mathcal{F}_T = \mathcal{G}_T = \mathcal{M}(T^*)$ and $\mathcal{Th}(\mathcal{E}_T) = \mathcal{Th}(\mathcal{F}_T) = T^f = \mathcal{Th}(\mathcal{G}_T) = \mathcal{Th}(\mathcal{A}_T) = T^F$ is the deductive closure of T*;

(2) the following are equivalent:

(a) T has the joint embedding property,

(b) T^f is a complete theory,

(c) T^F is a complete theory.

The following chart summarizes the current state of knowledge of the classes $\mathcal{E}_T$, $\mathcal{F}_T$, $\mathcal{G}_T$, and $\mathcal{A}_T$ for various theories T.

TABLE I

Model-Companions and Forcing Classes for Algebra

Theory	Status of $\mathcal{TA}(\mathcal{E}_T)$	Subclasses of Σ_T
Fields	Model-completion	$\mathcal{E}_T = \mathcal{F}_T = \mathcal{G}_T$ = class of algebraically closed fields; $\mathcal{A}_T$ = class of universal domains.
Ordered fields	Model-completion	$\mathcal{E}_T = \mathcal{F}_T = \mathcal{G}_T$ = class of ordered, real closed fields; $\mathcal{A}_T$ = class of ordered, real closed fields which include all Dedekind cuts defined by elements of the real closure of finitely generated subfields.
Formally real fields	Model-companion	$\mathcal{E}_T = \mathcal{F}_T = \mathcal{G}_T$ = class of real closed fields; $\mathcal{A}_T$ = class of real closed fields which include all Dedekind cuts defined by elements of the real closure of finitely generated subfields.
Nonarchimedean valued fields	Model-completion	$\mathcal{E}_T = \mathcal{F}_T = \mathcal{G}_T$ = class of algebraically closed, nonarchimedean valued fields whose value group is a densely ordered, divisible group.
Nonarchimedean valued fields with discretely ordered value group	Model-completion	$\mathcal{E}_T = \mathcal{F}_T = \mathcal{G}_T$ = the class of Hensel fields (J. Ax and S. Kochen (4)).

TABLE I (continued)

Model-Companions and Forcing Classes for Algebra

Theory	Status of $\mathcal{TA}(\mathcal{E}_T)$	Subclasses of Σ_T
Differential fields of characteristic 0	Model-completion	
Differential fields of characteristic p	Model-companion	
Abelian groups	Model-completion	(see P. Eklof and G. Sabbagh (34))
Torsion-free abelian groups	Model-completion	$\mathcal{E}_T = \mathcal{F}_T = \mathcal{G}_T$ = class of divisible, torsion-free abelian groups; $\mathcal{A}_T$ = class of divisible, torsion-free abelian groups which are infinite dimensional as a vector space over the rationals.
Groups	Not a model-companion	$\Sigma_T = \mathcal{P}_T$, $\mathcal{F}_T \cap \mathcal{G}_T = \emptyset$ (A. Macintyre (56)). $\mathcal{E}_T \supsetneqq \mathcal{E}_2 \supsetneqq \mathcal{E}_3 \supsetneqq \cdots \supsetneqq \mathcal{G}_T \supsetneqq \mathcal{A}_T$.
Metabelian groups	Not a model-companion	$\mathcal{F}_T \cap \mathcal{G}_T = \emptyset$ (D. Saracino (96)).
n-solvable groups*	Not a model-companion	(D. Saracino (96)).
Commutative rings	Not a model-companion	$\mathcal{F}_T \cap \mathcal{G}_T = \emptyset$ (G. Cherlin (17)).

* Announcements of results for nilpotent groups and Lie algebras have appeared in references 112-114.

TABLE I (continued)

Model-Companions and Forcing Classes for Algebra

Theory	Status of $\mathcal{TA}(\mathcal{E}_T)$	Subclasses of Σ_T
Commutative rings without nilpotent elements	Model-companion	(D. Saracino and L. Lipshitz (98)).
Commutative p-rings	Model-completion	(G. Sabbagh (91)).
Boolean algebras	Model-completion	$\mathcal{E}_T = \mathcal{G}_T = \mathcal{A}_T$ = class of atomless Boolean algebras.
R-modules for a coherent ring	Model-completion	(P. Eklof and G. Sabbagh (34)).
R-modules for a noncoherent ring	Not a model-companion	$\mathcal{E}_T = \mathcal{G}_T = \mathcal{A}_T$.
Division algebras	Not a model-companion	$\Sigma_T = \mathcal{P}_T$; $\mathcal{F}_T \cap \mathcal{G}_T = \emptyset$; $\mathcal{E}_T \supsetneq \mathcal{E}_2 \supsetneq \mathcal{E}_3 \supsetneq \cdots \supsetneq \mathcal{G}_T \supsetneq \mathcal{A}_T$.
Number theory	Not a model-companion	$\mathcal{E}_T = \mathcal{P}_T \cap \mathcal{M}(T_{\forall\exists})$; $\mathcal{F}_T = \{N\}$; $\mathcal{F}_T \cap \mathcal{G}_T = \emptyset$; $\mathcal{E}_T \supsetneq \mathcal{E}_2 \supsetneq \mathcal{E}_3 \supsetneq \cdots \supsetneq \mathcal{G}_T \supsetneq \mathcal{A}_T$.
Peano arithmetic	Not a model-companion	

PART TWO

ARITHMETIC

As was pointed out by A. Robinson [85] the interest in existentially complete models for arithmetic arises from the fact that while having an interesting algebraic structure they are still models of an important fragment of arithmetic. At an early stage of our work a third feature of these models showed itself; there is a close connection between some classes of existentially complete models and classes of second order ω-models of (weak) arithmetic. For example - the generic models correspond naturally to the ω-models of full second order arithmetic.

In Chapter 8 we show that all existentially complete models are models of the Π_2 fragment of arithmetic, that all the notions from recursion theory are naturally interpreted in such models and that existential completeness coincides with closedness under partial recursive functions. The last result yields the notion of existential closure.

In existentially complete models the set N of natural numbers may be defined by a formula. In the last section of Chapter 8 we investigate the best possible definitions of N: An $\exists\forall$ formula without parameters and a universal formula with (an arbitrary non-standard) parameter.

Having the notion of existential closure it seems that the simplest existentially complete models are those which are the closure of a single element. We call such models simple models and discuss them briefly in Chapter 9. We show how they may be constructed similarly to ultrapowers, using partial recursive functions and r.e sets only. An easy but interesting result is that every countable model of arithmetic can be embedded in a simple model.

Next we consider regular models. These are models that do not include cofinal simple models. (We do not have an example of a model which is neither simple nor regular). The algebraic structure of the regular models is investigated in Chapter 10. One of the results is that regular models are determined by the collection of subsets of N which are (existentially) definable in them. This leads to Chapter 11 and to a study of the relations between regular models and the corresponding second order models (the correspondence is one to one for countable models). It is shown how each of the two kinds of models can be interpreted in terms of the other. Here the generic models are exactly those which correspond to elementary submodels of the standard second order model (and all such substructures are obtained), and existentially universal models are those which correspond to the standard second order model (Chapter 12). Since the correspondence is computable (for countable models) we prove that the forcing companion of arithmetic is not an analytic set.

We do not have an interesting characterization of the class of second order models which are obtained from regular models. The following modification however, yields an interesting result: A model is biregular if its infinite part does not include a coinitial infinite part of a simple model. Biregularity implies regularity and although it looks like an algebraic property, expressible by a simple sentence we have the following: The countable biregular models correspond effectively in a one to one manner to all the countable ω-models of arithmetical comprehension (Chapter 11).

Most of Chapter 12 is devoted to the study of the approximating chains for arithmetic (see Chapter 5). It is shown that $\mathcal{E}_n$ consists of the models which correspond to (weak) β_n-models and that the theory of $\mathcal{E}_n$ is a complete Π^1_n set. The analogues for the $\mathcal{K}$-chain is briefly discussed. In particular we prove that all the approximating chains are strictly decreasing.

Finally, in Chapter 13 we apply the ideas of Chapter 8 to determine the finitely generic models for any complete theory T which includes Peano's axioms. It is shown that such a theory has a unique finitely generic model which is the existential closure of N in any model of T. It is worth mentioning that although this gives a clue how to handle the finite forcing companion of Peano's theory P, the question what are P^f and P^F is not yet answered.

Previous work on forcing and arithmetic was done by A. Robinson [85] and By Macintyre, Simmons and Goldrei [63]. Robinson was the first one to prove that the set of natural numbers is definable in existentially complete models and Macintyre, Simmons and Goldrei showed that the forcing companion of arithmetic is not an analytic set.

We mention also M.O. Rabin's paper [75] where it was first shown that non standard models of arithmetic are not existentially complete. In this paper Rabin introduced a method of proof which was modified by H. Friedman [38] and was used here often (Chapter 10).

Finally we remark that the key to the whole subject lies in the theorem of Y.V. Matijasevic [65] which identifies r.e predicates with existential formulas. As observed already by H. Gaifman [39] this relates recursion theory with model theoretic aspects of arithmet

CHAPTER 8

EXISTENTIALLY COMPLETE MODELS

§ 1 Models of T_{π_2}

We begin our discussion by investigating a fragment of Arithmetic which will be shown to hold in every existentially complete model. The language that we use has the function symbols $+$ and $\cdot$, the constant symbols 0 and 1 and the relation symbol $<$. Bounded quantifiers are introduced as

$$\exists x{<}t\phi \equiv \exists x(x < t \wedge \phi)$$

$$\forall x{<}t\phi \equiv \forall x(x < t \to \phi)$$

We shall use also the abbreviation $\exists\bar{x}{<}t\phi$ for $\exists x_1{<}t\cdots\exists x_n{<}t\phi$.

A bounded formula is a formula which has only bounded quantifiers. If ϕ is bounded then $\exists\bar{x}\phi$ is called a $\sum_1$ formula and $\forall\bar{x}\phi$ is called a Π_1 formula. More generally we define by induction: if ϕ is a $\sum_n$ formula and ψ is a Π_n formula, then $\forall\bar{x}\phi$ is a Π_{n+1} formula and $\exists\bar{x}\psi$ is a $\sum_{n+1}$ formula.

Let $T = Th(N)$ be the complete theory of the natural numbers N. The fragment that interests us is T_{π_2} - all Π_2 statements that hold in N.

8.1 Every model of T_{π_2} is linearly ordered by $<$ and includes N as an initial segment. N is the standard model of T_{π_2} and all other models are non-standard. If $M \models T_{\pi_2}$ then the elements of N are standard or finite and the elements of $M-N$ are non-standard or infinite.

8.2 First we adapt the notions of recursion theory to our framework. We observe that if $\phi(\bar{x},y)$ is in $\sum_1$ and $\psi(\bar{x})$ and $\chi(\bar{x})$ are in $\sum_1 \cup \Pi_1$ then the following have Π_2 normal form:

i) $\psi(\bar{x}) \leftrightarrow \chi(\bar{x})$,

ii) $\forall\bar{x}\forall y\forall z(\phi(\bar{x},y) \wedge \phi(\bar{x},z) \to y = z)$,

iii) $\forall\bar{x}\exists y\phi(\bar{x},y)$

Thus if ψ and χ as above describe the same relation on the natural numbers then they do so in every model of T_{π_2}. If $\phi(\bar{x},y)$ defines a partial function or a total function on N, then it does so in every model of T_{π_2}.

8,3 This justifies the following definitions which agree with the comm ones on N [e.g., Shoenfield (103)].

Definitions:

i) A formula which is equivalent in T_{π_2} to a Σ_1 formula is also called an r.e formula.

ii) A formula which is equivalent in T_{π_2} to a Σ_1 formula and to a Π_1 formula is called recursive.

iii) Let $\phi(\bar{x},y)$ be an r.e formula. $\phi(\bar{x},y)$ describes a partial recursive function if

$$T_{\pi_2} \vdash \forall \bar{x} \forall y \forall z [\phi(\bar{x},y) \wedge \phi(\bar{x},z) \rightarrow y = z].$$

iv) If in addition $T_{\pi_2} \vdash \forall \bar{x} \exists y \phi(\bar{x},y)$, then $\phi(\bar{x},y)$ describes a (total) recursive function.

The difference between our definitions and the standard ones is that we excluded some formulas which are equivalent to Σ_1 formulas in T but not in T_{π_2}. From our point of view they just "happen" to describe an r.e. predicate in N. Dealing with T_{π_2} we must stick to the simplest descriptions like the Σ_1 formulas themselves.

8.4 In every model M of T_{π_2} a formula $\phi(\bar{x},y)$ as above actually describes a (partial) function which, to be accurate, should be denoted by f_ϕ^M. We shall however refer freely to "the (partial) function f" having in mind a Σ_1 formula $\phi(\bar{x},y)$ which describes it and we shall write "$f(\bar{x}) = y$" instead of $\phi(\bar{x},y)$. By 8.2 the choice of $\phi(\bar{x},y)$ does not matter. We shall speak of "$\psi(f(\bar{x}))$ - the substitution of $f(\bar{x})$ in the formula $\psi(z)$" meaning the formula $\exists y[\phi(\bar{x},y) \wedge \psi(y)]$ (this includes the notion of composition of functions). These notatior seem proper in view of the absoluteness of recursive functions in model of T_{π_2} (8.8) and of partial recursive functions in existentially complete models (8.17).

8.5 It follows from the definitions that the class of r.e formulas is closed under conjunctions, disjunctions and prefixing of existential quantifiers. It is also closed under substitution of partial recursive functions. In particular - partial recursive functions are closed under compositions.

The class of recursive formulas is closed under conjunctions, disjunctions and negations. It is also closed under substitutions of (total) recursive functions. To show this we assume that $\phi(x)$ is equivalent in T_{π_2} to the Σ_1 formula $\psi(x)$ and to the Π_1 formula $\chi(x)$. The function $f(\overline{z}) = y$ is given by the Σ_1 formula $\theta(\overline{z},y)$. Then $\phi(f(\overline{z}))$ is described by both formulas

$$\exists y[\theta(\overline{z},y) \wedge \psi(y)]$$

and $$\forall y[\theta(\overline{z},y) \to \chi(y)]$$

The first formula is Σ_1 and the second one is Π_1 so that $\phi(f(\overline{z}))$ is recursive.

8.6 The class of recursive formulas is closed under prefixing of bounded quantifiers.

Here we must overcome some difficulties that arise from the fact that, according to our definitions, the appropriate equivalences must hold in T_{π_2} and not just in N. Indeed, in 9.8 we show that (unlike in N) the claim is not true for r.e formulas.

<u>Proof.</u> Let ϕ be recursive, so that $T_{\pi_2} \vdash \phi \leftrightarrow \psi$ and $T_{\pi_2} \vdash \phi \leftrightarrow \chi$, where ψ is in Σ_1 and χ in Π_1. We show that $\forall x{<}t\phi$ is also recursive. Let $\psi = \exists\overline{y}\psi_1$ where ψ_1 is bounded. Then

8.6.1 $$T \vdash \forall x{<}t\chi \leftrightarrow \forall x{<}t\phi$$

$$T \vdash \forall x{<}t\phi \leftrightarrow \forall x{<}t\exists\overline{y}\psi_1$$

and $$T \vdash \forall x{<}t\exists\overline{y}\psi_1 \leftrightarrow \exists z\forall x{<}t\exists\overline{y}{<}z\psi_1$$

The last equivalence holds in N but not necessarily in T_{π_2}.

By transitivity we get

8.6.2 $\forall x<t\chi \leftrightarrow \exists z\forall x<t\exists\overline{y}<z\psi_1$

which together with 8.6.1 shows that $\forall x<t\phi$ is equivalent to the Π_1 formula $\forall x<t\chi$ and to the Σ_1 formula $\exists z\forall x<t\exists\overline{y}<z\psi_1$. As 8.6.1 and 8.6.2 in their normal form are Π_2, the proof is completed.

8.7 Matijasevic's theorem [65, 66] relates r.e formulas to existential formulas. In our terminology it states: If $\phi(\overline{x})$ is a Σ_1 formula, then there is an existential formula $\psi(\overline{x})$ such that

$$T \vdash \forall\overline{x}(\phi(\overline{x}) \leftrightarrow \psi(\overline{x})).$$

This statement has a Π_2 normal form and therefore belongs to T_{π_2}.

8.8 Thus among models of T_{π_2} r.e formulas persist under extensions. In other words: if $M \subset M_1$ are models of T_{π_2} and $\phi(\overline{x})$ is a Σ_1 formula such that $M \models \phi(\overline{a})$, then $M_1 \models \phi(\overline{a})$. If $f(\overline{x})$ is a partial recursive function and $f(\overline{a})$ is defined in M, then $f(\overline{a})$ is defined also in M_1 and has the same value.

Recursive formulas and functions are absolute in the class of mode of T_{π_2} (persist under extensions and substructures). To prove this we still must show that recursive functions can be described by recursive formulas. Let $\phi(\overline{x},y)$ be a Σ_1 formula which defines a total function. Then

$$T_{\pi_2} \vdash \phi(\overline{x},y) \leftrightarrow \forall z(\phi(\overline{x},z) \rightarrow z = y).$$

Clearly the right hand side is Π_1.

8.9 We refer now to one of the main tools that we shall use: Kleene's enumeration theorem [e.g; Rogers (87)].
In our terminology it says:

i) For each n there is an n+1-place Σ_1 formula $F^n(y,\overline{x})$ with the following property: If $\phi(\overline{x})$ is an n-place Σ_1 formula then for some i

8.9.1 $T \vdash \forall\overline{x}(\phi(\overline{x}) \leftrightarrow F^n(i,\overline{x}))$.

Furthermore (the fixed point theorem): If $\psi(\overline{x},y)$ is any n+1-place Σ_1 formula, then there is a monotone recursive function $f(y)$ such that

$$T \vdash \forall y \forall \bar{x}(\psi(y,\bar{x}) \leftrightarrow F^n(f(y),\bar{x})) .$$

ii) For every n there is a partial recursive function $V^n(y,\bar{x})$ with the following property: If $f(\bar{x})$ is a partial recursive function, then there is an i such that for every $\bar{x}$ $f(\bar{x}) = V^n(i,\bar{x})$ (in particular - the functions are defined for the same n-tuples).

Remarks: i) 8.9.1 holds in T_{π_2}, and by 8.7 we may also assume that $F^n(y,\bar{x})$ is existential.

ii) Part (ii) of the definition must be made precise using the Σ_1 formulas $\psi(z,\bar{x})$ and $\chi(z,y,\bar{x})$ which describe f and V^n respectively:

$$T \vdash \forall \bar{x} \forall z(\psi(z,\bar{x}) \leftrightarrow \chi(z,i,\bar{x}))$$

T may be replaced by T_{π_2} so that $f(\bar{x}) = V^n(i,\bar{x})$ holds in all models of T_{π_2}.

iii) When dealing only with a fixed n we may omit the superscript and write F or V instead of F^n or V^n.

8.10 The axiom of induction holds in models of T_{π_2} for recursive formulas, i.e., if $\phi(\bar{x},y)$ is recursive, then

$$T_{\pi_2} \vdash \forall \bar{x}[\exists y \phi(\bar{x},y) \rightarrow \exists y(\phi(\bar{x},y) \wedge \forall z<y \sim\phi(\bar{x},z))] .$$

This follows from 8.5 and 8.6: $\exists y \phi(\bar{x},y)$ is equivalent to a Σ_1 formula and so is also $\exists y(\phi(\bar{x},y) \wedge \forall z<y \sim\phi(\bar{x},z))$. Therefore the whole statement is Π_2.

8.11 An easy but important corollary is that if M is a non-standard model of T_{π_2} then N is not definable in M by a recursive formula, even with the use of parameters. In other words, if for some fixed $\bar{a} \in M$ and for all $i \in N$, $M \models \phi(\bar{a},i)$, where ϕ is recursive, then $M \models \phi(\bar{a},b)$ also for some infinite b. To see this one applies 8.10 to $\sim\phi(\bar{a},y)$.

8.12 We specify some recursive functions that we shall use

i) $f(i) = p_i$ the ith prime number.

ii) $f(x_o,\ldots,x_n) = 2^{x_o}\ldots p_n^{x_n}$.

iii) $f(x) = (x)_i$ the highest power of p_i that divides x.

We shall also use the recursive relations

$R(i,x) \equiv p_i|x$ p_i divides x

$R'(i,x) \equiv p_i\nmid x$ p_i does not divide x.

Any $\sum_1$ formula which describes such a function or relation on N extends it to every model of T_{π_2} independently of the choice of the formula (8.2).

8.13 Although sets described by $\sum_1$ formulas with parameters may fail to have minimal elements (see 8.29) we can still relate to such formulas partial recursive Skolem functions: If $\phi(\overline{x},y)$ is a $\sum_1$ formula, then there is a partial recursive function $f(\overline{x})$ such that

8.13.1 $T_{\pi_2} \models \forall\overline{x}[\exists y\phi(\overline{x},y) \rightarrow \phi(\overline{x},f(\overline{x}))]$.

To obtain the function f, let $\phi(\overline{x},y)$ be $\exists\overline{z}\phi'(\overline{x},y,\overline{z})$ where ϕ' is bounded, and put

$$\psi(\overline{x},v) \equiv \phi'(\overline{x},(v)_o,(v)_1,\ldots(v)_n) \wedge \forall u<v \sim\phi'(\overline{x},(u)_o,(u)_1,\ldots(u)_n)$$

ψ is a recursive formula which describes the partial recursive function $h(\overline{x})$. It is easy to see that the definition $f(\overline{x}) = (h(\overline{x}))_o$ satisfies the requirements of 8.13.1.

§ 2 Existentially Complete Models for Arithmetic

We return now to the class $\sum_T$ of submodels of models of T, where T, as before, is the set of sentences which are true in N. We denote by $\mathcal{E}$ the class of existentially complete models in $\sum_T$

and by $T_{\mathcal{E}}$ the set of statements which are true in all models of $\mathcal{E}$.

8.14 We note that N itself is existentially complete; every element in N has a name in the language so that if $N \models \sim\phi(\bar{n})$ where ϕ is existential then $T \vdash \sim\phi(\bar{n})$ and $\phi(\bar{n})$ does not hold in any extension of N in Σ. Thus $T_{\mathcal{E}} \subset T$. Also, as T is complete, Σ_T has the joint embedding property.

8.15 Let $M' \subset M$. we say that M' is Σ_1 complete in M if for each Σ_1 formula $\phi(\bar{x})$ and each $\bar{a} \in M$

$$M \models \phi(\bar{a}) \text{ iff } M' \models \phi(\bar{a}).$$

The following theorem relates the subject of existentially complete models to recursion theory:

<u>Theorem.</u> If $M \models T_{\pi_2}$, $M' \subset M$ and M' is existentially complete in M, then M' is Σ_1 complete in M.

<u>Proof.</u> We show first that if $\phi(\bar{x},y)$ is a Σ_1 formula such that for some $\bar{a} \in M'$ and $b \in M$, $M \models \phi(\bar{a},b)$, then there is a $b' \in M'$ such that $M \models \phi(\bar{a},b')$. Assume that $M \models \phi(\bar{a},b)$. By Matijasevic's theorem (8.7) there is an existential formula $e(\bar{x},y)$ such that

8.15.1 $\quad T_{\pi_2} \vdash \forall \bar{x}y[\phi(\bar{x},y) \leftrightarrow e(\bar{x},y)]$

As $M \models \phi(\bar{a},b)$, we have $M \models e(\bar{a},b)$. M' is existentially complete in M, so $M \models e(\bar{a},b')$ for some $b' \in M'$. Applying 8.15.1 once more, we conclude that $M \models \phi(\bar{a},b')$.

Now we prove the theorem by induction on the number of quantifiers (both bounded and unbounded), assuming that every formula is in prenex normal form. If ϕ is atomic the statement is trivial.

Assume that the theorem holds for all Σ_1 formulas with at most n quantifiers. Let $\phi(\bar{a})$ be a Σ_1 formula with $n+1$ quantifiers. We assume first that $\phi(\bar{a}) = \exists x\psi(\bar{a},x)$ (or $\phi(\bar{a}) = \exists x{<}t\phi(\bar{a},x)$, which is similar as $\exists x{<}t\phi(\bar{a},x) = \exists x(\phi(\bar{a}) \wedge x{<}t)$).

If $M' \models \phi(\bar{a})$, then $M' \models \psi(\bar{a},b')$ for some $b' \in M'$ and by induction assumption $M \models \psi(\bar{a},b')$ so that $M \models \phi(\bar{a})$.

And, if $M \models \phi(\bar{a})$, then $M \models \psi(\bar{a},b)$ for some $b \in M$. By the first part of the proof $M \models \psi(\bar{a},b')$ for some $b' \in M'$ and again by induction assumption $M' \models \psi(\bar{a},b')$. Hence $M' \models \phi(\bar{a})$.

On the other hand, if the first quantifier is universal then by the definition of Σ_1 formulas ϕ must be bounded. Hence $\sim\phi$ (in normal form) is bounded, has n+1 quantifiers, and the first one is existential. Therefore the proof above applies to $\sim\phi$ and the claim for ϕ follows.

8.16 <u>Corollary.</u> If $E \in \mathcal{E}$ then $E \models T_{\pi_2}$. Equivalently $T_{\pi_2} \subset T_{\mathcal{E}}$

<u>Proof.</u> If $\phi \in T_{\pi_2}$ then $\phi = \forall\bar{x}\psi(\bar{x})$ where $\psi \in \Sigma_1$. Let E be existentially complete and M a model of T which extends E. Then for every $\bar{a} \in E$ $M \models \psi(\bar{a})$ and by 8.15 also $E \models \psi(\bar{a})$. Therefore $E \models \forall\bar{x}\psi(\bar{x})$.

8.17 Thus 8.1-8.13 hold also if we replace T_{π_2} by $T_{\mathcal{E}}$ and every Σ_1 formula can be thought of as an existential formula whenever we deal with existentially complete models. For such models the results of 8.8 are strenthened by 8.15; in $\mathcal{E}$ r.e formulas and partial recursive functions are absolute. If f is partial recursive and $f(\bar{a})$ is defined in a model $M \models T_{\pi_2}$ then $f(\bar{a})$ is defined also in every submodel that contains $\bar{a}$ and is existentially complete in M.

8.18 Another characterization of existential completeness is the following

<u>Theorem.</u> If $M \models T_{\pi_2}$ and $M' \subset M$, then M' is existentially complete in M iff it is closed under partial recursive functions (as a subset of M).

<u>Proof.</u> Immediate by 8.13 and 8.17.

8.19 Closedness under functions is preserved by intersections. Hence

<u>Corollary.</u> If $\{M_i | i \in I\}$ is a set of submodels of M which are existentially complete in M and $M \models T_{\pi_2}$, then $\bigcap\{M_i | i \in I\}$ is existentially complete in M.

8.20 Partial recursive functions are closed under compositions. Thus every subset (or submodel) of a model of T_{π_2} has an existential closure in that model, namely the closure of the set under partial recursive functions. By 8.18 this really yields the smallest submodel that includes the set and is existentially complete in that model. In particular if $E \in \mathcal{E}$, then the closure of any subset of E under partial recursive functions yields again a model in $\mathcal{E}$.

8.21 The remark above has some algebraic flavor, and it is natural to ask which models of T_{π_2} have an existential closure that is absolute, i.e., for what models do all existentially complete extensions yield the same existential closure (up to isomorphism)? The answer is disappointing form the point of view of Arithmetic. If $M \models T_{\pi_2}$ (or even $M \models T_{\forall\exists}$), then, unless it is already existentially complete, it can be extended to two models in $\mathcal{E}$ so that the existential closures are not isomorphic. We outline the proof:

a) If there is an existential formula with parameters in M that does not hold in M but holds in some extension, then there is such a formula $\phi(a)$ with a single parameter a. This we show by a careful use of the function $2^{x_0} \ldots p_n^{x_n}$ (careful to include the case where M is a model of $T_{\forall\exists}$ only).

b) The formula $\phi(x)$ describes an r.e set A in N. We decompose A into two r.e sets B, C such that

   i) $A = B \cup C$, $B \cap C = \phi$;

   ii) if D is an r.e set such that $D \cap B = \phi$ or $D \cap C = \phi$, then $D \cap {\sim}A$ is r.e.

   This decomposition is due to Friedberg (up to some small modifications) and the construction may be found in [(87), §12.2].

c) We choose existential formulas $\psi(x)$ and $\chi(x)$ describing B and C respectively and show that each of the formulas $\psi(a)$ and $\chi(a)$ is satisfied in some existentially complete extension of M.

d) On the other hand $\psi(a)$ and $\chi(a)$ cannot be satisfied simultaneously so that the closures of M in the two existentially complete models are not isomorphic over M.

The last result supplies an interesting example for the general theory of forcing in model theory. It shows:

Every pregeneric model of $T_{\forall\exists}$ is already existentially complete.

§ 3 The Definition of N in Existentially Complete Models

We turn next to one of the main features of existentially complete models which makes them very different from non-standard models of Arithmetic - the possibility of defining the natural numbers in such models. This result was first obtained by A. Robinson in [85] using a different method.

8.22 First we recall the notion of a simple set in recursion theory. An r.e set S is simple if its complement is an infinite set which does not contain any infinite r.e subset. Such sets exist [87] and from now on we fix one such set S and a single existential formula $s(x)$ which describes S in N.

8.23 Lemma. If $E \in \mathcal{E}$ and $a \in E-N$ then $E \models s(a)$.

Proof. Assume that $E \models \sim s(a)$. As E is existentially complete there is (by 1.6) an existential formula $\phi(x)$ such that

i) $E \models \phi(a)$

ii) $N \models \forall x(\phi(x) \rightarrow \sim s(x))$

From (ii) we conclude that the r.e set described by ϕ is disjoint from S and by the definition of simple sets it must be finite. Hence there are $n_1, \ldots, n_k \in N$ such that

$$N \models \forall x(\phi(x) \rightarrow x = n_1 \vee \ldots \vee x = n_k)$$

This statement holds also in E so that by (i) a is one of the n_i's and $a \in N$.

8.24 From now on we denote by $I(y)$ the $\exists\forall$ formula

$$\exists x[y<x \wedge \sim s(x)]$$

<u>Theorem.</u> If $E \in \mathcal{E}$ and $a \in E$ then

$E \models I(a)$ iff $a \in N$.

<u>Proof.</u> By Lemma 8.23 iff $E \models I(a)$ then $a \in N$. Assume on the other hand that $a \in N$. As the complement of S is infinite there is a $b \in N$ such that $a < b$ and $N \models \sim s(b)$. N is existentially complete (8.14) so that

$$E \models a < b \wedge \sim s(b) \ ;$$

in particular, $E \models I(a)$.

8.25 a) Thus the natural numbers are defined in every existentially complete model by an $\exists\forall$ formula without parameters. With the absence of parameters this is the best possible result. It is easy to see that if a Π_2 formula $\phi(x)$ holds for every element of N in some extension of N which is also a model of T_{π_2} then $N \models \forall x \phi(x)$. But then also $M \models \forall x \phi(x)$ whenever $M \models T_{\pi_2}$.

b) The sentence $\forall x I(x)$ is an $\forall\exists\forall$ sentence which holds in T but not in $T_{\mathcal{E}}$. A simple checking shows that this is the best possible result, as $\exists\forall\exists$ sentences of T become $\forall\exists$ after substituting a numeral.

c) The formula $\sim I(x)$ is an $\forall\exists$ formula which describes in every existentially complete model a set with no minimal element. Again it is easy to see that this is the strongest possible failure of the axiom of induction in models of T_{π_2} (for formulas without parameters).

8.26 Theorem 8.24 sheds some light on the relations between T and $T_{\mathcal{E}}$. Clearly $T \supset T_{\mathcal{E}} \cup \{\forall x I(x)\}$. On the other hand, if ϕ is a theorem of T, then ϕ^I - the relativization of ϕ to the predicate $I(x)$ is a theorem of $T_{\mathcal{E}}$, by 8.24. However in every model of $\forall x I(x)$ ϕ and ϕ^I state the same thing. Thus

$$T \equiv T_{\mathcal{E}} \cup \{\forall x I(x)\}.$$

8.27 To discuss the analogues of 8.24 and 8.25 for formulas with parameters and for further reference, we investigate more closely the way in which the formula $I(x)$ was constructed. Let $s(x)$ be as defined in 8.27. Replacing $\exists\bar{y}$ by $\exists y_{n+1}\exists y_1<y_{n+1}\cdots\exists y_n<y_{n+1}$ we assume that $s(x) = \exists y s'(x,y)$ where $s'(x,y)$ is bounded. We fix now two more formulas which are bounded:

$$p(x,y) = s'(x,y) \wedge \forall z<y \sim s'(x,z)$$

$$q(x,y) = p(x,y) \wedge \forall z<x \sim p(z,y)$$

These formulas describe partial recursive functions f_p and f_q with the following properties (which follow from the definitions):

i) $f_p(x)$ is defined for every non-standard element in every existentially complete model (8.23 and 8.10). f_q is a 1-1 partial function.

ii) $f_p \circ f_q(a) = a$ whenever $f_q(a)$ is defined.

$f_q \circ f_p(a) \leq a$ whenever $f_p(a)$ is defined.

iii) As N is existentially complete,

$f_p(N) \subset N$ and $f_q(N) \subset N$.

From this and (ii) we have also

$f_q(E-N) \subset E-N$.

8.28 The following theorem shows some more differences between existentially complete models and models of Arithmetic.

<u>Theorem.</u> Let E be existentially complete and let $a \in E-N$ be arbitrary.

i) The image of the initial segment determined by a under the partial function f_p is cofinal in E.

ii) The image of the terminal segment determined by a under f_q is coinitial in $E-N$.

Proof. (i) Assume on the contrary that it is bouned by b. Then by the definition of f_p (8.27)

8.28.1 $E \models \forall x{<}a[\exists y s'(x,y) \to \exists y{<}b s'(x,y)]$

We claim that

$$x \in N \text{ iff } \exists y{<}a[x{<}y \wedge \forall z{<}b \sim s'(y,z)]$$

If $m \in N$ then for some n we have $m < n$ and $T \models \forall z \sim s'(n,z)$. Therefore the statement holds with $y = n$.

On the other hand, if the statement holds for $x = c$ then there is an element d such that $c < d < a$ and $E \models \forall z{<}b \sim s'(d,z)$. By 8.28.1 $E \models \forall z \sim s(d,z)$ and $d \in N$ by 8.23. As $c < d$ we have also $c \in N$.

Thus we have defined N by a bounded formula, which is impossible by 8.11.

ii) Let $b \in E-N$ be given. Applying part (i) with b instead of a we get an element c such that $c < b$ and $f_p(c) = d > a$. But then $f_q(d) < c < b$ by 8.27 (ii). To complete the proof it remains to show that the image of the terminal segment is contained in $E-N$, and this is easy (8.27 (iii)).

8.29 Theorem. There is a Π_1 (and therefore a universal) formula $\phi(x,y)$ such that for every model $E \in \mathcal{E}$ and every element $a \in E-N$

$$N = \{x \in E | E \models \phi(x,a)\} .$$

Proof. Define $\phi(x,y) \equiv \forall z{<}x[\exists v s'(x,v) \to \exists v{<}y s'(x,v)]$. Clearly this is a Π_1 formula. We show that $E \models \phi(c,a)$ iff $c \in N$.

Assume that $c \in N$. If $b < c$ and $E \models \exists v s'(b,v)$, then $N \models \exists v s'(b,v)$ so that $\phi(c,a)$ holds.

On the other hand if $x \in E-N$ then by 8.28 for some $b < a$ $f_p(b) > a$, i.e., $\exists v s'(b,v)$ and no such v is smaller than a.

8.30 Thus $\phi(x,y)$ is a $\sum_1$ formula for which the induction axiom (in the version of minimal element) does not hold in non-standard existentially complete models. We do not know whether the induction axiom hold in existentially complete models for Π_1 formulas. We can however strengthen 8.11 for models in $\mathcal{E}$ and show that 8.29 is the best possible definition of N in existentially complete models.

<u>Theorem.</u> If $E \in \mathcal{E}$ is a non-standard model and $\psi(\bar{x},y)$ is an existential formula such that for some $\bar{a} \in E$ and for all $n \in N$, $E \models \psi(\bar{a},n)$ then $E \models \psi(\bar{a},b)$ also for some infinite b.

<u>Proof.</u> Assume that $E \models \psi(\bar{a},n)$ for all $n \in N$. By the compactness theorem there is an extension M of E in which $\psi(\bar{a},b)$ holds also for some $b \in M-N$. We extend M to some existentially complete model E' and choose an element $c \in E-N$. Let $\phi(x,y)$ be the formula that defines N as in 8.29. Then $E' \models \psi(\bar{a},b) \wedge \sim\phi(b,c)$. This is an existential formula so that for some $b' \in E$ $E \models \psi(\bar{a},b') \wedge \sim\phi(b',c)$.

CHAPTER 9

SIMPLE MODELS AND R.E ULTRAPOWERS

In this chapter we discuss briefly the simplest non-standard existentially complete models - those which are generated by a single non-standard element (to be more precise - by a single existential type). We show that a construction similar to that of ultrapowers, using only r.e sets and partial recursive functions yields all the simple models.

9.1 If E is existentially complete and $c \in E$ we denote by $N_E[c]$ the model E' which is the existential closure of c in E - the closure under partial recursive functions (8.20).
We note for further reference that, with the definitions of V and I as in 8.9 and 8.24, we have

$$a \in E' \text{ iff } E \models \exists y[I(y) \wedge V^1(y,c) = a] .$$

As partial recursive functions are absolute in existentially complete models we have $E' = N_{E'}[c]$. Thus we define

<u>Definition:</u> A model $E \in \mathcal{E}$ is <u>simple</u> if for some $c \in E$ $E = N_E[c]$, in which case we say that c is a generator for E and that E is generated by c.

9.2 We note that simple models are the existentially complete models which are also models of the sentence

$$\exists z \forall x \exists y[I(y) \wedge V^1(y,z) = x]$$

We note also that if E is finitely generated, $E = N_E[a_o,\ldots,a_n]$, then E is simple, as $E = N_E[b]$ where $b = 2^{a_o}\ldots p_n^{a_n}$. Thus for example every countable existentially complete model is the union of a chain of simple models.

9.3 The simple models may be identified with (realizations of) existential types. For this we need the following lemma:

Lemma. Let E and E' be existentially complete such that $E = N_E[a_1,\dots,a_n]$ and $E' = N_{E'}[b_1,\dots,b_n]$ and such that $\tau(a_1,\dots,a_n) = \tau(b_1,\dots,b_n)$ ($\bar{a}$ and $\bar{b}$ realize the same existential type), then there is a (unique) isomorphism between E and E' which maps a_i to b_i, $i = 1\dots n$.

Proof. Clearly the only possible way of obtaining an isomorphism F is to define, for every partial recursive function g,

$$F: g(a_1,\dots,a_n) \to g(b_1,\dots,b_n) \quad ;$$

and it is easy to see that all the properties of an isomorphism (well-defined, 1-1, onto, and preserves operations) correspond to existential properties of $\bar{a}$ and $\bar{b}$ which are the same.

Thus, if Φ is any existential 1-type., we define $N[\Phi]$ to be the simple model which is generated by an element that has the type Φ; i.e., if E is any model in which an element c realizes Φ, then $N[\Phi] = N_E[c]$. By the preceding lemma the definition does not depend on the choice of E and c.

9.4 Next we describe a standard procedure for obtaining all simple models. We construct models similar to ultrapowers using only part of the functions and part of the sets.

An r.e filter is a set of r.e sets which has the finite intersection property. An r.e ultrafilter is a maximal r.e filter. Every r.e ultrafilter may be identified with an existential type by replacing every r.e set in it by an existential formula which describes that set. More precisely, if F is an r.e ultrafilter, then the following is an existential type:

$$\tau(F) = \{\phi(x) \mid \phi \text{ is existential and } \{x \mid N \models \phi(x)\} \in F\}.$$

On the other hand, if Φ is an existential type, then the following is an r.e ultrafilter:

$$\mathrm{Fil}(\Phi) = \{A \subset N \mid \text{there is a formula } \phi \in \Phi \text{ such that } A = \{y \mid N \models \phi(y)\}\}$$

This correspondence is one-to-one and onto, and it is easy to check that

$$\mathrm{Fil}(\tau(F)) = F , \qquad \tau(\mathrm{Fil}(\Phi)) = \Phi$$

whenever F is an r.e ultrafilter and Φ an existential 1-type.

Let C be the set of partial recursive functions and F be an r.e ultrafilter. We define an equivalence relation on the subset of C of functions whose domain is in F:

$$f \equiv_F g \ \text{ iff } \ \{x \mid f(x) = g(x)\} \in F.$$

It is easy to see that this is an equivalence relation, and we denote by f/F or by $[f]$ the equivalence class of the function f.

The r.e ultrapower C/F is the structure whose universe is the set of equivalence classes. 0 and 1 are the (classes of the) functions which are 0 and 1 identically. Addition and multiplication are defined by

$$[f] + [g] = [f+g] \quad , \quad [f][g] = [fg] ,$$

and $[f] < [g]$ iff $\{x \mid f(x) < g(x)\} \in F$.

Again it is easy to see that the operations are well defined.

9.5 The fact that r.e ultrapowers are just the simple models may be summarized by the following theorem, whose proof follows straightforwardly from the definitions.

<u>Theorem.</u> Let F be an r.e ultrafilter and Φ an existential 1-type.

i) C/F is isomorphic to $N[\tau(F)]$.

ii) $N[\Phi]$ is isomorphic to $C/\mathrm{Fil}(\Phi)$.

In both cases the isomorphism is obtained by identifying the generator of the simple model with (the class of) the identity function.

9.6 The construction above emphasizes again the relations between recursion theory and existentially complete models, and we feel that these relations deserve a more thorough investigation. We mention some results on simple models that follow from the correspondence above.

i) There are minimal non-standard existentially complete models (N is their only existentially complete submodel).

ii) There are simple models which are generated by a single eleme[nt] and the (total) recursive functions, but not all simple models have this property.

iii) Simple models do not have non-trivial automorphisms (we do not know if they may be isomorphic to proper submodels).

To prove (i) we construct an r.e ultrafilter such that every partial recursive function is either undefined or constant or one-to-one on some set in the filter.

For (ii) we construct an r.e ultrafilter that has a base consisting of recursive sets. On the other hand an ultrafilter which contains a maximal r.e set B and all r.e sets which include the complement of B yields a model without this property.

The proof of (iii) is based on the lemma that if F is an r.e ultrafilter and f is a one-to-one partial recursive function such that $f(A) \in F$ whenever $A \in F$ then f is the identity on some set in F. This is shown by decomposing the set on which f differs from the identity function into three r.e. sets such that $f(A_i) \cap A_i = \phi$ for $i = 1,2,3$.

9.7 The simple models may be very complicated in their structure. In fact we have

<u>Theorem.</u> Every countable model of T_{π_2} can be embedded in a simple model.

<u>Proof.</u> Let M be such a model. We order M in a sequence $\langle a_1 \dots a_n \dots \rangle_{n<\omega}$, add a new constant c and look at the theory

$$T' = \mathit{Th}(M) \cup \{(c)_i = a_i \mid i < \omega\}$$

where $(c)_i$ is the projection defined in 8.12.

T' is consistent, as any finite subtheory can be interpreted in M, with the proper choice of c. Therefore there is an elementar[y] extension M' of M which is a model of T'. We extend M' to an existentially complete model E and finally look at the simple model $E' = N_E[c]$. As all models described above are models of T_{π_2}, the recursive functions preserve their values so that for every $i < \omega$ $(c)_i = a_i$ in E'. Hence $E' \supset M$.

9.8 To conclude this chapter we use simple models to prove that r.e formulas are not closed in T_{π_2} under the prefixing of bounded quantifiers.

Let $\phi(x,z,v)$ be the $\sum_1$ formula $\exists y<z[V^1(y,v) = x]$. We show that $\forall x<v\phi(x,z,v)$ is not equivalent in T_{π_2} to a $\sum_1$ formula. By 9.2 every simple model is a model of $\exists v \exists z<v \ \forall x<v\phi(x,z,v)$ (choose for v a generator and for z any non-standard element smaller than v).

If $\forall x<v\phi(x,z,v)$ were equivalent in T_{π_2} to a $\sum_1$ formula then $\exists v \exists z<v \forall x<v\phi(x,z,v)$ would be equivalent to a $\sum_1$ statement which, holding a simple models, would hold also in N. This however is impossible, as it asserts the existence of numbers n and m such that $m < n$ and such that applying the $m-1$ first partial recursive functions to n yields all the values $0,1,\ldots n-1$.

CHAPTER 10

REGULAR MODELS

10.1 Definition: A model $E \in \mathcal{E}$ is regular if, for every $a \in E$, $N[a]$ is bounded in E.

10.2 We note that the definition is elementary: The predicate

$$B(x,y) \equiv \forall z[I(z) \rightarrow V(z,x) < y]$$

states that $N[x]$ is bounded by y. Thus a model $E \in \mathcal{E}$ is regular iff $E \models \forall x \exists y B(x,y)$. (Strictly speaking, we must replace $V(z,x) < y$ by $\forall t(V(z,x) = t \rightarrow t < y)$, as V is only a partial function).

Every $M \in \sum_T$ can be easily extended to a regular model. By comptness we extend M to M' which has an element bigger than all members of M. We then extend M' to an existentially complete model E'. Repeating this construction ω times and taking the union, we get an existentially complete model and for every element which is added at stage n, $N[c]$ is bounded at stage $n+1$.

10.3 Definition. Let E be existentially complete and let A be a set of natural numbers. The set A is existentially defined in E if there is an existential formula $e(y,\bar{x})$ and an n-tuple $\bar{a} \in E$ such that

$$A = \{y \in N \mid E \models e(y,\bar{a})\} .$$

We say that the formula $e(y,\bar{a})$ defines A, and we denote by S_E the collection of sets in $P(N)$ which are existentially defined in

Note that, as we allow only definitions by existential formulas the definitions are persistent: if $e(y,\bar{a})$ defines A in E and $E \subset E'$ then the same formula also defines A in E'.

For regular models we observe the following:

If E is regular and $A \subset N$ is existentially defined in E, then A is also defined by a bounded formula. The reason for this is that if A is defined by the formula $\exists \bar{z}\, e(y,\bar{a},\bar{z})$ where e is quantifier-free and if $b \in E$ is a bound for $N[\bar{a}]$, then A is defined by the formula $\exists \bar{z} < b\, e(y,\bar{a},\bar{z})$, since for any standard i if $\exists \bar{z}\; e(i,\bar{a},\bar{z})$ has a solution in E then it has one in $N[\bar{a}]$.

Actually we can choose the definition of any existentially definable set in a very uniform way:

10.4 Lemma. Let E be regular and $A \subset N$ a set which is defined in E by the bounded formula $\phi(i)$ (with parameters in E). Then there is an element $c \in E-N$ such that

$$A = \{i \in N \mid E \models P_i | c\} .$$

Proof. Define the predicate

$$\psi(j) \equiv \exists y[j < y \wedge \forall i < j(\phi(i) \longleftrightarrow P_i|y)] .$$

Clearly $\psi(j)$ holds for all $j \in N$, and by 8.6 $\psi(j)$ is a $\sum_1$ formula. By 8.30 $\psi(j_o)$ holds also for some $j_o \in E-N$, and there is a $y_o \in E$ such that $E \models \forall i < j_o(\phi(i) \leftrightarrow P_i|y_o)$.

Note that choosing some $d \in E-N$ and writing $\exists y < c$ instead of $\exists y$ in $\psi(j)$ we can make y_o arbitrary small in $E - N$.

The proof of lemma 10.4 reveals the main idea that will be used throughout this section. In 10.5 the same method is used to prove a basic result on regular models. We recall that $F(i,\bar{x})$ is the enumeration of the existential formulas as defined in 8.9.

10.5 Lemma. Let E be regular, $\bar{a} \in E$ and $\tau(\bar{a})$ the existential type of $\bar{a}$. Let $\Phi(\bar{x},y)$ be an existential type which extends $\tau(\bar{a})$ such that

$$A = \{i \in N \mid F^{n+1}(i,\bar{x},y) \in \Phi(\bar{x},y)\}$$

is existentially definable in E. Then there is an element $b \in E$ such that $\tau(\bar{a},b) = \Phi$.

Proof. Let $\phi(x)$ be the bounded formula which defines A in E and let c be a bound for $N[\bar{a}]$ in E. Let $F^{n+1}(i,\bar{x},y)$ be $\exists \bar{v}\; F(i,\bar{x},y,\bar{v})$ where F is quantifier-free. We define the formula

$$\psi(j) \equiv \exists y \,\forall i < j[\phi(i) \to \exists \bar{v} < c\; F(i,\bar{a},y,\bar{v})] .$$

ψ is a $\sum_1$ formula. We show first that $E \models \psi(j)$ for all $j \in N$. Let $j \in N$ be given and let $i_1,\ldots,i_r$ be the elements smaller than j for which $\phi(i)$ -holds in E. Thus the following formula is in $\Phi(\overline{x},y)$:

$$\exists \overline{v}\ F(i_1,\overline{x},y,\overline{v}) \wedge \ldots \wedge \exists \overline{v}\ F(i_r,\overline{x},y,\overline{v}).$$

As $\Phi(\overline{x},y)$ extends $\tau(\overline{a})$

$$N[\overline{a}] \models \exists y[\exists \overline{v}\ F(i_1,\overline{a},y,\overline{v}) \wedge \ldots \wedge \exists \overline{v}\ F(i_r,\overline{a},y,\overline{v})]\ .$$

As c is a bound for $N[\overline{a}]$ in E we get $E \models \psi(j)$.
Thus $E \models \psi(j_o)$ also for some $j_o \in E\text{-}N$, and there is a $b \in E$ such that for all $i \in N$, $E \models \phi(i) \rightarrow F^{n+1}(i,\overline{a},b)$. It is now immediate that $\Phi(\overline{x},y) \subset \tau(\overline{a},b)$ as F^{n+1} enumerates all existential formulas, and equality holds as $\Phi(\overline{x},y)$ is a maximal consistent set of existential formulas.

Using lemma 10.5 we prove first that for regular models S_E completely determines E.

10.6 Theorem.

i) If E and E' are regular and $S_E = S_{E'}$, then $E \equiv E'$ in $L_{\infty,\omega}$.

ii) If, in addition, $\tau(\overline{a}) = \tau(\overline{b})$ for some n-tuples $\overline{a} \in E$ and $\overline{b} \in E'$, then $E \equiv E'$ in $L_{\infty,\omega}(c_1,\ldots,c_n)$ where the new constants are interpreted as $\overline{a}$ in E and as $\overline{b}$ in E'.

iii) If, in addition, $E \subset E'$, then $E \prec E'$ in $L_{\infty,\omega}$.

Proof. i), ii) By Karp's criterion [51] it is enough to show that there exists a family J of monomorphisms of submodels in E onto submodels in E' such that given $f \in J$ and $a \in E$ there is a monomorphism g extending f with a in its domain. Similarly: given $b \in E'$ f can be extended to include b in its range.

If $\tau(\overline{a}) = \tau(\overline{b})$, then $N[\overline{a}]$ is isomorphic to $N[\overline{b}]$. We define J to be the family of isomorphisms between simple models extending $N[\overline{a}]$ in E and simple models extending $N[\overline{b}]$ in E', carrying $\overline{a}$ to $\overline{b}$.

Let $f:N[\bar{a},\bar{c}] \to N[\bar{b},\bar{d}]$ be an isomorphism in J and let c' be any element of E. As f is an isomorphism, $\tau(\bar{a},\bar{c}) = \tau(\bar{b},\bar{d})$. Thus, $\tau(\bar{a},\bar{c},c')$ extends $\tau(\bar{b},\bar{d})$, and by lemma 10.5 there is an element $d' \in E$ such that

$$\tau(\bar{a},\bar{c},c') = \tau(\bar{b},\bar{d},d') .$$

By 9.3 there is a (unique) isomorphism g between $N[\bar{a},\bar{c},c']$ and $N[\bar{b},\bar{d},d']$ which extends f and carries c' to d'. A similar argument shows that the range of an isomorphism can be extended to include a given element.

Part iii follows from part ii. To show that $E \prec E'$ in $L_{\infty,\omega}$ we must show that $E \equiv E'$ in $L_{\infty,\omega}(\underline{c}_1,\ldots,\underline{c}_n)$ for any finite addition of constants naming elements of E. Letting $\bar{a} = \bar{c}$ and $\bar{b} = \bar{c}$ in part ii we get $E \equiv E'$ in $L_{\infty,\omega}(\underline{c}_1,\ldots,\underline{c}_n)$. Thus $E \prec E'$ in $L_{\infty,\omega}$.

By Scott's theorem [52] there is an $L_{\omega_1,\omega}$ sentence for any given countable structures such that all its countable models are isomorphic. Thus we have

10.7 <u>Corollary.</u> If E and E' are regular and countable such that $S_E = S_{E'}$, then E and E' are isomorphic.

Moreover: If $\bar{a} \in E$ and $\bar{b} \in E'$ realize the same existential n-type then the isomorphism can be chosen to carry $\bar{a}$ to $\bar{b}$.

In particular, the assumptions of corollary 10.7 are satisfied if we let E equal to E' and choose $\bar{a}$ and $\bar{b}$ to be n-tuples realizing the same existential n-type. This yields another important property of regular models:

10.8 <u>Corollary.</u> If E is countable and regular and if $\bar{a}$ and $\bar{b}$ realize in E the same existential n-type, then there is an automorphism carrying $\bar{a}$ to $\bar{b}$.

If E is uncountable we still obtain a very strong result: If $\bar{a}$ and $\bar{b}$ realize the same existential n-type, then $E \equiv E$ in $L_{\infty,\omega}(\underline{c}_1,\ldots,\underline{c}_n)$ interpreting $\underline{c}_1 \ldots \underline{c}_n$ as $\bar{a}$ on one side and $\bar{b}$ on the other. Thus, we have also

10.9 <u>Corollary.</u> If E is regular and $\tau(\bar{a}) = \tau(\bar{b})$ in E, then for any formula $\phi(\bar{x})$ in $L_{\infty,\omega}$ $E \models \phi(\bar{a})$ iff $E \models \phi(\bar{b})$.

A. Robinson proved in [81] a general theorem in model theory which is related to Corollary 10.9: If M is a generic model for some theory K, then the existential type of an n-tuple in M determines its complete type (in $L_{\omega,\omega}$) (see Corollary 3.15).

We turn next to another property of regular models which deals with the possibility of defining elements by formulas (theorem 10.11).

10.10 <u>Lemma.</u> Let E be regular and $\bar{a}, b \in E$. If $b \notin N[\bar{a}]$ then there is an element $b' \in E$ such that $b' < b$ and $\tau(\bar{a},b) = \tau(\bar{a},b')$.

<u>Proof</u>: We show first that there is no formula $\phi(\bar{x},y) \in \tau(\bar{a},b)$ such that b is the minimal solution of $\phi(\bar{a},y)$. Assume that $\phi(\bar{x},y)$ is such a formula. Then $\forall y{<}b \sim\phi(\bar{a},y)$ is universal. As E is existentially complete, there is an existential formula $e(\bar{x},y)$ such that $E \models e(\bar{a},b)$ and

10.10.1 $$T \vdash \forall \bar{x},z[e(\bar{x},z) \rightarrow \forall y{<}z \sim\phi(\bar{x},y)]$$

Therefore $E \models \phi(\bar{a},b) \wedge e(\bar{a},b)$, and as $N[\bar{a}]$ is existentially complete there is an element $c \in N[\bar{a}]$ such that $E \models \phi(\bar{a},c) \wedge e(\bar{a},c)$. From 10.10.1 it now follows that c is the minimal solution of $\phi(\bar{a},y)$ in E, so that $c = b$ and $b \in N[\bar{a}]$. But this contradicts the hypothesis of the lemma. Thus, we can assume that any existential formula having $\bar{a}$ as parameters which is satisfied by b is also satisfied by an element smaller than b.

Let c be a bound for $N[\bar{a},b]$ and let $F(i,\bar{x},y) = \exists \bar{z} F'(i,\bar{x},y,\bar{z})$ where F' is quantifier-free. Let $A = \{i \in N \mid E \models F(i,\bar{a},b)\}$ and let $\phi(i)$ be a bounded formula which defines A. Put

$$\psi(j) \equiv \exists y{<}b\, \forall i{<}j[\phi(i) \rightarrow \exists \bar{z}{<}c F'(i,\bar{a},y,\bar{z})]$$

$\psi(j)$ is a bounded formula. For any finite j let $i_1,\ldots,i_r$ be the elements smaller than j for which $\phi(i)$ holds. Then

$$E \models F(i_1,\bar{a},b) \wedge \ldots \wedge F(i_r,\bar{a},b)$$

By the first part of the proof $E \models$ $y{<}b(F(i_1,\bar{a},y)$ ... $F(i_r,\bar{a},y))$.
As $N[\bar{a},b]$ is existentially complete in E,

$$N[\bar{a},b] \models \exists y{<}b(\exists \bar{z} F'(i,\bar{a},y,\bar{z}) \wedge \ldots \wedge \exists \bar{z} F'(i_r,\bar{a},y,\bar{z})) ,$$

so that $E \models \exists y{<}b(\exists \bar{z}{<}c F'(i,\bar{a},y,\bar{z}) \wedge \ldots \wedge \exists \bar{z} F'(i_r,\bar{a},y,\bar{z}))$.

Thus $\psi(j)$ holds for every finite j and therefore also for some infinite j_o which yields a b' such that

$$b' < b \wedge \forall i<j_o[\phi(i) \rightarrow \exists \bar{z}<cF'(i,\bar{a},y,\bar{z})].$$

It is now immediate that $\tau(\bar{a},b) = \tau(\bar{a},b')$.

10.11. Theorem. If E is regular and $\phi(\bar{x},y)$ is a formula in $L_{\infty,\omega}$ such that for some $\bar{a}$, $b \in E$ b is the unique solution in E $\phi(\bar{a},y)$, then $b \in N[\bar{a}]$.

Proof: By 10.10 if $b \notin N[\bar{a}]$ then there is an element $b' \in E$ such that $\tau(\bar{a},b) = \tau(\bar{a},b')$ and $b \neq b'$. By 10.9 we get $E \models \phi(\bar{a},b')$.

Thus if b can be defined in terms of $\bar{a}$ by an $L_{\infty,\omega}$ formula then it can be obtained from $\bar{a}$ by a partial recursive function.

In particular, if E is regular then the only elements definable in $L_{\infty,\omega}$ without parameters are the standard elements. For simple models the situation is completely different: If $E = N[c]$ then by 9.3 and 9.6, c is the only element which realizes $\tau(c)$ and generates the whole model. Thus c is defined by an $L_{\omega_1,\omega}$ formula and similarly also every other element (as the image of c under some particular partial recursive function).

Unlike simple models regular models have many automorphisms.

10.12. Theorem. Let E be regular and countable. For every $\bar{c} \in E$ the submodel which is invariant under all automorphisms that fix $\bar{c}$ is exactly $N[\bar{c}]$.

Proof. An automorphism that fixes $\bar{c}$ clearly fixes $N[\bar{c}]$. On the other hand, if $b \notin N[\bar{c}]$ then by 10.10 $\tau(\bar{c},b) = \tau(\bar{c},b')$ for some $b' < b$. By 10.8 there is an automorphism which carries $\bar{c}$ to itself and b to b'.

We mention another property of regular models.

10.13. Theorem. If E is regular and countable and if $c \in E-N$, then there is an elementary submodel of E which is bounded by c and is isomorphic to E.

Proof. To avoid repetitions (the idea is the same as in 10.5 and 10.10) we shall only outline the proof. We order E in a sequence $\langle a_1,\dots,a_n,\dots\rangle_{n<\omega}$ and construct by induction a sequence $\langle b_1,\dots,b_n,\dots\rangle_{n<\omega}$ such that

i) $\tau(a_1,\dots,a_n) = \tau(b_1,\dots,b_n)$,

ii) $N[b_1,\dots,b_n]$ is bounded by c.

The n+1 step is similar to the proof of 10.5 with one modificatio - we replace the formula $F^{n+1}(i,\bar{x},y)$ by the formula $\exists z F^{n+2}(i,\bar{x},y,z$ This will ensure that $N[b_1,\dots,b_{n+1}]$ is also bounded by c.

Incidently - the same modification strengthens Friedman's theore for models of T [38]: "If $M \models T$ and $c \in M-N$ then M is isomorphic to an initial segment of itself which is bounded by c". This is not true for models of Peano's arithmetic where c must satisfy a condition which is stronger than $c \in M-N$. The nature of this condition will be made clear in Chapter 13.

Biregular Models

The notion of biregularity is the dual of the notion of regularity. Biregular models will play an important role in Chapters 11 and 12.

10.14 Definition. Let E be an existentially complete model.

i) $b \in E$ is a lower bound for $\bar{c} \in E$ (or for $N[\bar{c}]$) if $b \in E-N$ and for every $d \in N[\bar{c}] - N$, $b < d$.

ii) E is biregular if every $c \in E$ has a lower bound.

10.15 Remark. The condition for biregularity (of an existentially complete model) can be written down in the first order language. The predicate "b is a lower bound for $\bar{c}$" can be expressed as

$$LB(\bar{c},b) \equiv \forall i,y[i \notin N \vee y \in N \vee \sim V^n(i,\bar{c}) = y \vee b<y] \wedge b \notin N$$

Using the $\forall\exists$ definition for $b \notin N$ (8.25(c)) and the universal definition of N with b as parameter (8.29) to write $i \notin N$ and $y \in N$, $LB(x,y)$ becomes an $\forall\exists$ formula. Note also that we denote by LB different formulas with different numbers of variables, depending on the length of $\bar{c}$.

10.16 Lemma. Every biregular model is regular.

Proof: Let $E \in \mathcal{E}$ be non regular. Then for some $c \in E$ $N[c]$ is cofinal in E. We shall show that $N[c]-N$ is coinitial in $E-N$. Let $b \in E-N$ be arbitrary (small). We fix an element $a \in N[c]-N$. By 8.26 and 8.27 there is an element $d \in E-N$ such that $a < d$ and $f_q(d) < b$. By the cofinality of $N[c]$ we can also choose $a' \in N[c]-N$ such that $d < a'$. Define

$$e(a,a',x) \equiv \exists y{<}a'[a < y \wedge f_q(y) = x].$$

$f_q(y) = x$ is the bounded formula $q(x,y)$ so that $e(a,a',x)$ is bounded and has a minimal solution x_0 in E, which must be also in $N[c]$ by 8.8. Clearly $x_0 \leqslant d' < b$ and $x_0 \notin N$ by 8.26(iii).

CHAPTER 11

REGULAR MODELS AND SECOND ORDER MODELS FOR ARITHMETIC

In this chapter we relate to every regular model a second order model of arithmetic. We investigate the relations between the theory of the regular model and that of the corresponding second order model. In section 2 we show that the biregular models correspond to (all) the second order models of arithmetical comprehension. As a corollary we prove that there are $2^{\aleph_0}$ pairwise non elementary equivalent biregular models.

For a given regular model E we defined in section 3 the set S_E of subsets of N which are defined in E by an existential formula. By lemma 10.4 any member of S_E is defined by the formula $P_i | x$ substituting some element of E for x. In this section, we shall treat N together with S_E as a model for second order arithmetic. Thus in the language L^2 we have two kinds of variables $xyz \ldots$ and $XYZ \ldots$. We have quantifiers $\exists x$ $\forall x$ and $\exists X$ $\forall X$ where xyz range over elements of N and XYZ over sets. Atomic formulas are formulas of the form $t + s = r$, $t \cdot s = r$, $t < s$ and $t \varepsilon X$. We assume that 0 and 1 are constants in the language. The pairs $\langle N, S_E \rangle$ as models in this language correspond to the ω-models introduced in [40] with one difference: $\langle N, S_E \rangle$ in general is not a model for the axiom of comprehension. We denote by $\mathfrak{N}_0$ the standard model $\langle N, P(N) \rangle$.

For every regular model E we denote by $\mathfrak{N}_E$ the second order model induced by E - $\langle N, S_E \rangle$.

We wish to determine the elementary theory of $\mathfrak{N}_E$ from that of E. To every formula ϕ of L^2 we relate the formula $\overline{\Phi}$ in the original language of induction:

If $\phi \equiv t_1 = t_2$	then	$\overline{\Phi} \equiv t_1 = t_2$
If $\phi \equiv x \varepsilon Y$	then	$\overline{\Phi} \equiv P_x | y$
If $\phi \equiv \exists x \psi$	then	$\overline{\Phi} = \exists x(I(x) \wedge \overline{\Psi})$
If $\phi \equiv \exists X \psi$	then	$\overline{\Phi} = \exists x(\sim I(x) \wedge \overline{\Psi})$

($I(x)$ is the formula which describes N in E (8.25)).
$\wedge, \vee$ and $\sim$ are transferred as usual and $\forall x$, $\forall X$ like $\sim\exists x\sim$ and $\sim\exists X\sim$. By an easy induction we now obtain:

11.1 Lemma. Let E be regular and let $\phi(\overline{X},\overline{y})$ be a formula in L^2. For every $\overline{n} \in N$ and $\overline{A} \in S_E$

$$\mathcal{N}_E \models \phi(\overline{A},\overline{n}) \text{ iff } E \models \overline{\phi(a,n)}$$

where $a_1 \ldots a_n$ are elements in $E - N$ such that

$$A_k = \{i \in N \mid P_i | a_k\} \qquad k = 1 \ldots n.$$

Thus we also have

11.2 Theorem: Let E and E' be regular models.

i) For any sentence $\phi \in L^2$, $\phi \in T(\mathcal{N}_E)$ iff $\overline{\Phi} \in T(E)$.

ii) If $E \prec E'$ then $\mathcal{N}_E \prec \mathcal{N}_{E'}$

iii) If $E \equiv E'$ then $\mathcal{N}_E \equiv \mathcal{N}_{E'}$

iv) If $E \simeq E'$ then $\mathcal{N}_E = \mathcal{N}_{E'}$.

Next we show how to obtain the theory of a regular model E from that of $\mathcal{N}_E$. The relation is less obvious than in 11.1. The basic idea is to express the existential types of E in $\mathcal{N}_E$ and use the fact that the existential type determines the complete type.

We define the following formulas in L^2 using some easy abbreviations.

$$C^n(X) = \forall j \; \exists \overline{x} \; \forall i{<}j[i \in X \rightarrow F^n(i,\overline{x})]$$

$$\tau^n(X) = C^n(X) \wedge \forall i[i \notin X \rightarrow \sim C^n(X \cup \{i\})].$$

$$E^n_x(X,Y) = \tau^n(X) \wedge \tau^{n+1}(Y) \wedge \forall i[i \in X \rightarrow f^n(i) \in Y]$$

where f^n is the recursive function for which

$$N \models F^n(i,\overline{x}) \leftrightarrow F^{n+1}(f(i),\overline{x},x_{n+1}) \quad \text{(see 8.9)}$$

Thus $\tau^n(X)$ means that $\{F^n(i,\overline{x}) \mid i \in X\}$ is an existential n-type and $E^n_x(X,Y)$ actually corresponds to extension of types.

$\tau^o(X)$ holds for exactly one set - the set of indices of existential sentences true in N. We denote this set by τ_o. For every regular model E, $\tau_o \in S_E$.

If E is a regular model then the definitions above actually correspond in $\mathfrak{N}_E$ to the types in E. To prove this we need some more notations:

Let E be regular and $\mathfrak{N}_E$ the second order model induced by E. For a given n-tuple $\bar{a} \varepsilon E$ we denote by $A_{\bar{a}}$ the set

$$A_{\bar{a}} = \{i \varepsilon N \mid E \models F^n(i,\bar{a})\}$$

On the other hand, if $A \varepsilon \mathfrak{N}_E$ and $\mathfrak{N}_E \models \tau^n(A)$ then

$$\tau_A = \{F^n(i,\bar{x}) \mid i \varepsilon A\}$$

is actually an existential type.

By lemmas 10.4 and 10.5 we get immediately.

11.3 <u>Lemma:</u> Let E be a regular model and let $\mathfrak{N}_E$ be the second order model induced by E.

i) If $\bar{a}, b \varepsilon E$ then $A_{\bar{a}}$ and $A_{\bar{a},b}$ are in $\mathfrak{N}_E$ and

$$\mathfrak{N}_E \models E^n_x(A_{\bar{a}}, A_{\bar{a},b})$$

ii) If $A \varepsilon \mathfrak{N}_E$ and $\mathfrak{N}_E \models \tau^n(A)$ then there is an n-tuple $\bar{a} \varepsilon E$ such that $A = A_{\bar{a}}$. More generally

iii) If $\mathfrak{N}_E \models E_x(A_{\bar{a}}, B)$ then there is a $b \varepsilon E$ such that $B = A_{\bar{a},b}$.

To define the satisfaction function, we assume that ϕ is a formula of first order arithmetic in prenex normal form with m quantifiers and n free variables. We define the predicate by induction on m only:

$$m = 0 \quad Sat_\phi(X) \equiv \tau^n(X) \wedge \exists i \varepsilon X \ \forall x_1 \ldots x_n[\phi(\bar{x}) \leftrightarrow F^n(i,\bar{x})]$$

Namely the atomic formula is in the type.

For $m = k + 1$

$$\text{if } \phi = \exists x\psi \quad Sat_\phi(X) = \tau^n(X) \wedge \exists Y[E^n_x(X,Y) \wedge Sat_\psi(Y)]$$

$$\text{if } \phi = \forall x\psi \quad Sat_\phi(X) = \tau^n(X) \wedge \forall Y[E^n_x(X,Y) \rightarrow Sat_\psi(Y)].$$

Clearly the definition can be made explicitly for any given ϕ.

11.4 Lemma: Let E be regular. For any formula $\phi(\overline{x})$ in prenex normal form

$$E \models \phi(\overline{a}) \quad \text{iff} \quad \mathfrak{N}_E \models \mathrm{Sat}_\phi(A_{\overline{a}}) .$$

Proof: By induction on the number of quantifiers: The statement is true for $n = 0$. Assume ϕ has $n + 1$ quantifiers and $\phi = \exists x\psi(\overline{a},x)$. If $E \models \exists x\psi(\overline{a},x)$ then $E \models \psi(\overline{a},b)$ for some $b \in E$. By induction assumption $\mathfrak{N}_E \models \mathrm{Sat}_\psi(A_{\overline{a},b})$. By lemma 11.3 $\mathfrak{N}_E \models E_x(A_{\overline{a}},A_{\overline{a},b})$ and by the definition of Sat, $\mathfrak{N}_E \models \mathrm{Sat}_\phi(A_{\overline{a}})$. If on the other hand, $\mathfrak{N}_E \models \mathrm{Sat}_\phi(A_{\overline{a}})$. then $\mathfrak{N}_E \models \mathrm{Sat}_\psi(B)$ and $\mathfrak{N}_E \models E_x(A_{\overline{a}},B)$ for some $B \in \mathfrak{N}_E$. Again by 11.3 we have $B = A_{\overline{a},b}$ for some $b \in E$. Thus, $\mathfrak{N}_E = \mathrm{Sat}_\psi(A_{\overline{a},b})$ and by induction assumption $E \models \psi(\overline{a},b)$ so that $E \models \phi(\overline{a})$.

The proof for universal quantifier now follows.

If ϕ is a sentence, we denote by Sat_ϕ the sentence $\mathrm{Sat}_\phi(\tau_o)$. We have now the converse of 11.2:

11.5 Theorem:

i) If E is regular and $\mathfrak{N}_E$ is the induced second order model, then for any sentence ϕ in first order logic

$$E \models \phi \quad \text{iff} \quad \mathfrak{N}_E \models \mathrm{Sat}_\phi .$$

If E and E' are regular then

ii) If $\mathfrak{N}_E \equiv \mathfrak{N}_{E'}$ then $E \equiv E'$

iii) If $\mathfrak{N}_E \prec \mathfrak{N}_{E'}$ and $E \subset E'$ then $E \prec E'$

iv) If $\mathfrak{N}_E \prec \mathfrak{N}_{E'}$ and E is countable then there is an elementary embedding of E into E'.

Proof: i)-iii) follow immediately from 11.4. iv) follows from the observation that whenever E is countable and $S_E \subseteq S_{E'}$ an easy induction construction of an existential type preserving map yields a monomorphism of E into E'. By iii) the monomorphism is elementary.

Biregular Models and Models of Arithmetical Comprehension

It is natural to try to characterize the second order models which are obtained from regular models. We do not have a satisfactory answer to this question as our characterization includes reference to the notion of an existential type (or uses the predicates $C^n(X)$ and $\tau^n(X)$ which were previously defined). For biregular models 11.13 is a very interesting characterization. The proof is not easy and is divided to some lemmas and theorems.

11.6 Lemma: i) Let E be biregular and $e(\bar{x},y)$ a bounded formula with parameters $a_1,\dots,a_n \in E\text{-}N$. Let b be a lower bound for $N[a_1 \dots a_n]$ (see 10.14). Then for any $\bar{x} \in N$ $E \models \exists y(I(y) \wedge e(\bar{x},y))$ iff $E \models \exists y{<}b\ e(\bar{x},y)$. Similarly $E \models \forall y[I(y) \rightarrow e(\bar{x},y)]$ iff $E \models \forall y{<}b\ e(\bar{x},y)$. ($I(y)$ - the predicate which defines the natural numbers in 8.25) ii) More generally: Let $\phi(\bar{x}) = Q_1 z_1 \dots Q_k z_k\ e(\bar{x},\bar{z})$ be a formula with parameters $a_1,\dots,a_n \in E$ such that $e(\bar{x},\bar{z})$ is bounded and for every $i \leq k$ Q_i is a quantifier relativized to $I(X)$. Let $b_1 \dots b_k \in E\text{-}N$ be elements such that b_k is a lower bound for $N[a_1,\dots,a_n]$ and such that for every $i < k$ b_i is a lower bound for $N[a_1 \dots a_n\ b_{i+1} \dots b_k]$. Then there is a bounded formula $h(\bar{x},\bar{b})$ such that for all $\bar{x} \in N$

$$E \models \phi(\bar{x}) \quad \text{iff} \quad E \models h(\bar{x},\bar{b}).$$

Proof: i) Fix $\bar{x} \in N$. Clearly, if $E \models \exists y[I(y) \wedge e(\bar{x},y)]$ then $E \models \exists y{<}b\ e(\bar{x},y)$. If on the other hand, $E \models \exists y{<}b\ e(\bar{x},y)$ then there is such a minimal y_o by 8.10 and $y_o \in N[a_1,\dots,a_n]$. As $y_o < b$ $y_o \notin E\text{-}N$ so that $y_o \in N$ and $I(y_o)$ holds. The dual now follows taking negations.

ii) This follows from i) by a trivial induction on k.

The statement that b is a lower bound for $N[a_1 \dots a_n]$ is elementary (10.14) and therefore the transfer from relativized quantifiers to bounded ones can be done uniformly. More precisely:

11.7 Lemma: Let $\phi(x_1 \ldots x_m\, y_1 \ldots y_n) = Q_1 z_1 \ldots Q_k z_k\, e(\bar{x},\bar{y},\bar{z})$ be a formula such that $Q_i z_i$ are quantifiers relativized to $I(x)$ and such that e is a bounded formula. For every $i \leq k$ we denote by $\overline{Q_i z_i}$ the quantifier $\exists z_i < v_i$ if Q_i is existential and $\forall z_i < v_i$ if Q_i is universal. Let ϕ^* be the formula

$$\phi^*(x_1,\ldots,x_n,y_1,\ldots,y_n) = \exists v_1 \ldots v_k [\bigwedge_{i=1}^{k} LB(y_1 \ldots y_m, v_{i+1} \ldots v_k; v_i) \wedge$$

$$\wedge\, \overline{Q_1 z_1} \ldots \overline{Q_k z_k}\;\; e(\bar{x},\bar{y},\bar{z})]$$

Then for any biregular model E and every $\bar{x} \in N$ and $\bar{y} \in E\text{-}N$

$$E \models \phi(\bar{x},\bar{y}) \quad \text{iff} \quad E \models \phi^*(\bar{x},\bar{y}).$$

Note also that by 10.14 and 8.7 we can assume that ϕ^* is an $\exists\forall\exists$ formula.

A formula in the second order language L^2 is arithmetical if all its quantifiers are of the form $\exists x$ and $\forall x$.

11.8 Definition: A set $S \subset P(N)$ is closed under arithmetical definability if for any arithmetical formula ϕ with parameters in S and one free variable x

$$\{x \mid N \models \phi(x)\} \in S.$$

In this case we say also that S is a model of arithmetical comprehension.

11.9 Theorem: If E is biregular, then S_E is closed under arithmetical definability.

Proof: Let $\phi(A_1 \ldots A_n, x)$ be given such that $A_j \in S_E$ for $j = 1 \ldots n$. Let $a_1 \ldots a_n \in E\text{-}N$ be elements such that $A_j = \{i \in N \mid E \models P_i | a_j\}$ $j = 1 \ldots n$. Then $B = \{x | N \models \phi(\bar{A},x)\} = \{x | \mathcal{N}_E \models \phi(\bar{A},x)\} = \{x \in N | E \models \overline{\phi(\bar{a},x)}\}$. ($\overline{\Phi}$ was defined at the beginning of this chapter) $\overline{\Phi}$ may be assumed to be of the form $Q_1 z_1 \ldots Q_k z_k \psi$ where ψ is a bounded formula (note that although $P_x | y$ is not bounded it may be easily replaced by a bounded formula in which y is the bound for all quantifiers). By lemma 11.7 there is a bounded formula $h(\bar{a},\bar{b},x)$ such that

$$B = \{x \in N \mid E \models h(\bar{a},\bar{b},x)\}$$

so that $B \in S_E$.

To prove the converse, we need the following lemma:

11.10 Lemma: Let $S \subset P(N)$ be closed under arithmetical definabilit
i) If $A \in S$ and for some n $\{F^n(i,\bar{x}) \mid i \in A\}$ is consistent with T then there is a set B, $A \subset B \in S$ such that $\{F^n(i,\bar{x}) \mid i \in B\}$ is consistent with T, and maximal with respect to this property.
ii) If $\Phi(x_1 \ldots x_n)$ is an existential type such that $\{i \mid F^n(i,\bar{x}) \in \Phi\} \in S$ then any subset of N which is existentially definable in the simple model $N[\Phi]$ (see chapter 9) is in S.

Proof: i) Let A as above be given. We define by induction on n the predicate $\psi(n)$:

11.10.1 $\psi(n) \equiv \forall j \exists \bar{x}[\forall i<j(i \in A \to F(i,\bar{x})) \wedge \forall i<n(\psi(i) \to F(i,\bar{x})) \wedge F(n,\bar{x})]$.

As is well known $\psi(x)$ can be defined explicitly by an arithmetical formula so that $B \in S$ where $A \subset B = \{n \mid \mathfrak{N}_0 \models \psi(n)\}$. As $T = T(N)$ it is clear from 11.10.1 that $\{F^n(i,\bar{x}) \mid i \in B\}$ is consistent with T and that for $j \notin B$ $\{F^n(i,\bar{x}) \mid i \in B\} \cup \{F^n(j,\bar{x})\}$ is not consistent with

ii) Let $N[\Phi]$ be $N[a_1,\ldots,a_n]$ so that $B = \{i \in N \mid N[a] \models F^n(i,\bar{a})\}$ is in S. Let $A \subset N$ be a set which is existentially defined in $N[\bar{a}]$. Every element of $N[\bar{a}]$ is the image of $\bar{a}$ under some partial recursive function and we can assume that $A = \{i \in N \mid N[\bar{a}] \models e(i,\bar{a})\}$ where e is existential. By 8.9 there is a recursive function f such that

$$T_{\pi_2} \vdash \forall y\bar{x}[e(y,\bar{x}) \leftrightarrow F(f(y),\bar{x})]$$

Thus $A = \{i \mid f(i) \in B\}$ and as this is an arithmetical definition in terms of B we have also $A \in S$.

11.11 Theorem: If E is regular and if S_E is closed under arithmetical definability, then E is biregular.

Proof: Let c be an arbitrary element in E. We shall prove that $N[c]$ has a lower bound. Clearly we can assume $c \notin N$. Put $A = \{i \in N \mid E \models \exists z\, F^2(i,c,z)\}$ so that $A \in S_E$. We define the second order predicate

$$\psi(n) \equiv \forall j\, \exists x\, \exists y\, \forall i<j[(i \in A \to \exists z F^2(i,x,z)) \wedge j<y \wedge F^2(n,x,y)].$$

Claim: $\mathfrak{N}_o \models \psi(n)$ iff $F^2(n,c,y)$ is satisfied in $N[c]$ by some non-standard y.

Assume first $\mathfrak{N}_o \models \psi(n)$. Then

$$T_\forall \cup \{\exists z F(i,a,z) \mid i \in A\} \cup \{m < b \mid m \in N\} \cup \{F(n,a,b)\}$$

is consistent, and has a model E' which can be assumed to be existentially complete. As A enumerates the type $\tau(c)$ we can assume that $E \supset N[c]$, and $c = a$.
Thus $E' \models \exists y[y \notin N \wedge F(n,c,y)]$. The property $y \notin N$ can be expressed by an existential formula with c as parameter (8.29) so that the whole statement is existential. As N[c] is existentially complete

$$N[c] \models \exists y[y \notin N \wedge F(n,c,y)].$$

Conversely: Assume that $N[c] = F(n,c,b)$ for some $b \notin N$. For any given $j \in N$, let $i_1 \ldots i_r$ be the elements less then j in A. Then

$$N[c] = \exists x \exists y[\exists z F(i,x,z) \wedge \ldots \wedge \exists z F(i_r,x,z) \wedge j < y \wedge F(n,x,y)].$$

(as $x = c$ and $y = b$ satisfy this formula). But $N[c] \models T_{\pi_2}$ so that the same statement must hold also in N. As j was arbitrary $\mathfrak{N}_o \models \psi(n)$. This proves the claim.

We next put $B = \{n \mid \mathfrak{N}_o \models \psi(n)\}$. By the claim above the following is consistent

$$T_o = T_\forall \cup \{\exists z F(i,c,z) \mid i \in A\} \cup \{m < b_o \mid m \in N\} \{\exists y(F(n,c,y)\ b_o<y) \mid n \in B\}$$

(and indeed for any finite subset of T_o there is a choice of b_o which makes N[c] a model). Let E' be any existentially complete model of T_o. Then b_o is a lower bound for N[c] as any image d of c under a partial recursive function is the unique solution of the predicate F(i,c,y) for some i, and d is infinite iff $i \in B$ so that $b_o < d$.

To complete the proof we show that E itself can be turned into a model of T_o by an appropriate choice of b_o. Let f,g and h be the recursive function of 8.9 for which

11.11.1 $T_{\pi_2} \vdash \forall xyi[\exists z F^2(i,x,z) \leftrightarrow F^2(f(i),x,y)]$

$T_{\pi_2} \vdash \forall xyi[i < y \leftrightarrow F^2(g(i),x,y)]$

$T_{\pi_2} \vdash \forall xyi[\exists z(F^2(i,x,z) \wedge y < z) \leftrightarrow F^2(h(i),x,y)]$

Let $C = f(A) \cup g(N) \cup h(B)$. Then C is arithmetically defined with A and B as parameters so that $C \in S_E$. T_o is consistent and hence by 11.11.1 $T_\forall \cup \{F^2(i,x,y) \mid i \in C\}$ is consistent. By lemma 11.10 there is a set $D \in S_E$ such that $C \subset D$ and $\{F^2(i,x,y) \mid i \in D\}$ is maximal set which is consistent with $T_\forall$. Thus $\Phi = \{F^2(i,x,y) \mid i \in D\}$ is an existential type which extends $\tau(c)$. As $D \in S_E$ we get by lemma 10.5 an element $b \in E$ such that $\tau(c,b) = \Phi$. By the construction of Φ it is now clear that $E \models T_o$ and b is a lower bound for $N[c]$.

Thus E is biregular iff E is regular and S_E is a model of arithmetical comprehension. For countable models we have in addition:

11.12 <u>Theorem:</u> Let $S \subset P(N)$ be a countable collection which is closed under arithmetical definability. Then there is a biregular model E such that $S = S_E$.

<u>Proof:</u> Let S as above be given. In view of 11.11 it is enough to show that $S = S_E$ for some regular model E. We enumerate S: $S = \langle A_n \rangle_{n<\omega}$ and construct by induction an increasing sequence of simple models with the following properties:

i) E_n is bounded in E_{n+1}.

ii) A_n is existentially definable in E_n.

iii) All existentially definable sets in E_n are in S.

The conditions hold if we put $E_o = N$ (only condition iii) has a meaning). Assume E_n is given and $E_n = N[c]$. Let f, g and h be the recursive functions of 8.9 such that:

$$T \vdash P_i | y \leftrightarrow F^2(f(i),x,y)$$

$$T \vdash P_i \nmid y \leftrightarrow F^2(g(i),x,y)$$

$$T \vdash F^1(i,x) \leftrightarrow F^2(h(i),x,y)$$

Let $C = \{i \in N \mid N[c] \models F^1(i,c)\}$. By induction assumption $C \in S$. Thus the following set A is also in S $A = f(A_{n+1}) \cup g(A_{n+1}) \cup h(C)$. Clearly $\{F^2(i,x,y) \mid i \in A\}$ is consistent and by lemma 11.10 there is a binary existential type $\Phi(x,y)$ such that $B = \{i \in N \mid F^2(i,x,y) \in \Phi\}$ and $A \subset B \in S$. Let $N[a,b]$ be the simple model $N[\Phi]$. Clearly, $N[a]$ is isomorphic to $N[c]$ and A_{n+1} is defined in $N[a,b]$ by

$$A_{n+1} = \{i \in N \mid N[a,b] \models P_i | b\}.$$

As $B \in S$ we conclude by 11.10 that every existentially definable set in $N[a,b]$ is in S.

To assure that E_n is bounded in E_{n+1}, we repeat the construction above: Let g be the recursive function such that

$$T \vdash \forall xyzi[F^2(i,x,y) \wedge \exists t(V^2(i,x,y) = t \wedge t < z) \leftrightarrow F^3(g(i),x,y,z)]$$

Then $\{F^2(j,x,y,z) \mid j \in g(B)\}$ is consistent and can be extended to a trenary type $\Phi(x,y,z)$ such that in $N[\Phi] = N[a',b',c']$ there is no existentially defined set which is not in S. $N[a']$ is isomorphic to $N[a] \approx N[c]$ and is bounded in $N[\Phi]$ by c' and $N[a'b']$ is isomorphic to $N[a,b]$. Thus $E_{n+1} = N[\Phi]$ is a model as required.

Next we define $E = \bigcup\{E_n \mid n < \omega\}$. Then E is existentially complete as the union of a chain of existentially complete models. E is regular as for every $b \in E$ we have $b \in E_n$ for some n so that $N[b]$ is bounded in E_{n+1}. If $A \in S$ then A is existentially definable in E_n for some n, so that it is also definable in E and if $A \notin S$ then it is not existentially definable in any E_n and therefore not in E.

11.13 <u>Corollary:</u> The correspondence $E \to \mathcal{N}_E$ is 1-1 from all countable biregular models onto all countable models of arithmetical comprehension.

The class of arithmetical sets is a minimal model for arithmetic comprehension. Thus, we have by 11.5.

11.14 <u>Corollary:</u> There is a minimal biregular model E_o (for which S_E is the class of arithmetical sets) such that any biregular model includes E_o.

In [40] A. Grezegorzyk, A. Mostowski and C. Ryll Nardzewski defined the theory A. All models of A are models of arithmetical comprehension. Their undecidability theorem, combined with a Henkin type construction can be used to show that A has $2^{\aleph_0}$ complete extensions that have ω-models. This observation, together with 11.13 and 11.2 yield the following.

11.15 <u>Corollary:</u> There are $2^{\aleph_0}$ countable biregular models which are pairwise non elementary equivalent.

CHAPTER 12

GENERIC MODELS AND THE ANALYTIC HIERARCHY

§1 Generic and Existentially Universal Models

The relations between regular models and second order models established in chapter 11 yield a nice characterization of generic and existentially universal models in arithmetic:

i) E is existentially universal iff it is regular and $\mathfrak{N}_E$ is the standard second order model.

ii) E is generic iff it is regular and $\mathfrak{N}_E$ is an elementary substructure of the standard second order model. Moreover - the correspondence $E \to \mathfrak{N}_E$ relates all countable generic models to all countable elementary substructures of $\mathfrak{N}_o$.

A closer examination shows that under this correspondence classes which arise naturally in the research of model theory correspond to classes which arise in the research of Arithmetic. In particular we obtain in our theory complete Π^1_n sets.

We turn next to discuss the generic and existentially universal models for arithmetic. We recall that Σ_T is the class of submodels of models of arithmetic. $\mathcal{G}$ is the class of generic models, and T^F, the forcing companion, is the theory of the class $\mathcal{G}$. As T is complete T^F is also complete and every two generic models are elementarily equivalent.

We observe first the following:

12.1 Theorem: Every generic model is biregular.

Proof: Let G be a generic model and $a \in G$ an arbitrary element. By compactness we can extend G to a model M in which there is an element b which is infinite but smaller than the non-standard elements of G. This model can now be extended to a generic model

G' so that $G \prec G'$. As G is existentially complete $N_G[a] = N_{G'}[a]$ and

$$G' \models \text{"}b \text{ is a lower bound for } N[a]\text{"}.$$

The last statement is elementary (10.14) so that we have also

$$G \models \text{"there is an element that is a lower bound for } N[a]\text{"}$$

As, remarked any two generic structures are elementary equivalent. Thus, by 11.15 the class of biregular models is much bigger than $\mathcal{G}$.

In view of what was proved in chapter 11 the next easy theorem is the key to our results for generic structures.

12.2 Theorem: A model E is existentially universal iff it is regular and $S_E = P(N)$.

Proof: If E is existentially universal then it is regular by 12.1. Let $A \subset N$ be given then $T \cup D(E) \cup \{P_i|X \mid i \in A\} \cup \cup \{P_i \nmid X \mid i \notin A\}$ is consistent. Thus there is an element $c \in E$ such that $E \models P_i|c$ iff $i \in A$. Hence $A \in S_E$ on the other hand if E is regular and $S_E = P(N)$ then every existential type with (finitely many) parameters in E is realized in E by Lemma 10.4.

In particular we have:

i) Any existentially universal model has power $2^{\aleph_0}$ at least while there are $2^{\aleph_0}$ countable generic models pairwise non-isomorphic.

ii) A regular model that extends an existentially universal model is existentially universal.

Fix now an existentially universal model E. Then $\mathcal{Th}(E) = T^F$ as T^F is complete and $\mathcal{Th}(\mathfrak{N}_E) = \mathcal{Th}(\mathfrak{N}_0)$ where $\mathfrak{N}_0$ is the standard model. By 11.1 and 11.5:

If $\phi \in L^2$ then $\phi \in T(\mathfrak{N}_0)$ iff $\overline{\Phi} \in T^F$.

If $\psi \in L$ then $\psi \in T^F$ iff $Sat_\psi \in T(\mathfrak{N}_0)$.

We summarize:

12.3 Theorem:

i) T^F and $T(\mathfrak{N}_0)$ are recursively equivalent.

ii) T^F is not an analytic theory.

iii) If G is generic then it does not have an analytic diagram.

Proof:

i) We shall not call for the exact definitions that need some kind of Goedel's enumeration. It is clear that the mappings $\phi \to \overline{\phi}$ and $\psi \to Sat_\psi$ are effective and yield recursive reductions of $T(\mathfrak{N}_0)$ to T^F and vice versa.

ii) $T(\mathfrak{N}_0)$ is known to be non analytical so that by (i) the same is true for T^F.

iii) It is easy to check that if the diagram of G is analytical then so is also the complete theory of G. As $\mathcal{Th}(G) = T^F$, (iii) follows from (ii).

12.4 Theorem: The mapping $\sigma: E \to \mathfrak{N}_E$ maps the countable generic models onto the countable elementary submodels of $\mathfrak{N}_0$. It is one to one (up to isomorphism) and preserves (elementary) inclusions.

Proof: Every generic model G can be extended to an existentially universal model so that by 11.2 $\mathfrak{N}_G \prec \mathfrak{N}_0$. σ is 1-1 on the class of all countable regular models and every generic model is regular. Finally - every elementary submodel of $\mathfrak{N}_0$ is a model of arithmetical comprehension so that it is obtained from some regular model E (11.12). By 11.5 and 12.2 E can be elementarily embedded in an existentially universal model, so that E is generic. Thus σ is also onto.

§2 The Approximating Chains for $\mathcal{E}$

Modifying the definition of a β_n model we show that under the mapping $E \to \mathcal{N}_E$ the hierarchy of the $\mathcal{E}_n$ classes corresponds to the hierarchy of the β_n models.

12.5 Definition:

i) A formula of L^2 is Σ^1_0 or Π^1_0 if it is arithmetical (11.8)
A formula is in $\Sigma^1_{n+1}[\Pi^1_{n+1}]$ if it is of the form $\exists \bar{X}\phi(\bar{X})[\forall \bar{X}\phi(\bar{X})]$ where $\phi \in \Pi^1_n[\phi \in \Sigma^1_n]$.
A formula is analytic if it is in Σ^1_n for some n.

ii) If M and M' are in Σ we write $M \prec_n M'$ if any Σ_n formula with parameters in M, holds in M iff it holds in M'.

iii) If $\mathcal{N}$ and $\mathcal{N}'$ are second order models as defined at the beginning of chapter 11 we write $\mathcal{N} \prec^1_n \mathcal{N}'$ if any Σ^1_n formula with parameters in $\mathcal{N}$, holds in $\mathcal{N}$ iff it holds in $\mathcal{N}'$

iv) $\mathcal{N}$ is a β_n model if $\mathcal{N} \prec^1_n \mathcal{N}_o$.

Recall that $\mathcal{N}_o$ is the standard model and note that this definition differs from the usual one as we do not require that $\mathcal{N}$ will be a model of the analytic comprehension axiom.

The $\mathcal{E}$ hierarchy was defined in Chapter 4. We begin the discussion with the following observation:

12.6 Lemma: Every model in $\mathcal{E}_2$ is biregular.

Proof: By 12.1 every generic model is biregular. Let $E \in \mathcal{E}_2$ be given. We extend E to a generic model G. Let $c \in E$ be an infinite element. Then $N[c]$ has a lower bound in G. Therefore $G \models \exists x LB(c,x)$. By 10.15 this is a Σ_3 formula and hence by 4.2 $E \models \exists x LB(c,x)$.

12.7 Theorem: $E \in \mathcal{E}_{n+1}$ iff E is regular and $\mathcal{N}_E$ is a β_n model

Proof: We shall prove the theorem for the case $n = 1$. A similar proof takes care of the induction step in the general case.

Assume first that $E \in \mathcal{E}_2$. By 12.6 E is regular. Let G be an existentially universal model extending E. Thus $\mathcal{N}_G = \mathcal{N}_o$ and we must show that $\mathcal{N}_E \prec^1_1 \mathcal{N}_G$. Let $\phi(\bar{X},\bar{Y})$ be an arithmetical formula such that for some $\bar{A} \in \mathcal{N}_E$ $\mathcal{N}_G \models \exists \bar{X}\phi(\bar{X},\bar{A})$. Let a_j $j = 1 \ldots n$ be elements of E such that $A_j = \{i \in N |\ E \models P_i | a_j\}$. By 11.1 $G \models \overline{\exists \bar{X}\phi(\bar{X}, a_1 \ldots a_n)}$. By definition

$\overline{\exists \bar{X}\phi(\bar{X}, a_1 \ldots a_n)} = \exists x_1 \ldots x_m[\bigwedge_i (x_i \notin N) \wedge \psi(\bar{x},\bar{a})]$ where ψ has only quantifier relativized to $I(x)$ in front of a bounded formula. Thus for some m-tuple $\bar{b} \in G-N$ $G \models \psi^*(\bar{b},\bar{a})$ (see 11.7). $\psi^*(\bar{b},\bar{a})$ is an EAE formula and by 8.29 $\exists \bar{x}[\bar{x} \notin N \wedge \psi^*(\bar{x},\bar{a})]$ is also an EAE formula. As $E \prec_3 G$ (4.2) we have $E \models \exists \bar{x}[\bar{x} \notin N \wedge \psi^*(\bar{x},\bar{a})]$ so that $E \models \psi^*(\bar{c},\bar{a})$ for some $\bar{c} \in E-N$. E is also biregular so that $E \models \psi(\bar{c},\bar{a})$ and hence $E \models \overline{\exists \bar{x}\phi(\bar{x},\bar{a})}$. We conclude that $\mathcal{N}_E \models \exists \bar{x}\phi(\bar{X},\bar{A})$.

To show the converse we must prove that if $\mathcal{N}_E$ is a β_1 model then $E \prec_3 G$ for some generic model G. To this end we shall have to modify the definition of the predicate $Sat_\phi(A)$. As introduced in chapter 11 $Sat_\phi(A)$ is Σ^1_n if ϕ is Σ_n. By 1.2 we know that for any AE formula $\phi(\bar{x})$ and any existentially complete model E in which $\bar{a}$ realizes the type $\Phi(\bar{x})$ $E \models \phi(\bar{a})$ iff $T_\forall \cup \Phi(\bar{x}) \cup \phi(x)$ is consistent. Thus, we may define for an AE formula $\phi(\bar{x})$

$$Sat_\phi(A) \equiv \tau(A) \wedge \forall j \exists \bar{x}[\phi(\bar{x}) \wedge \forall i<j(i \in A \rightarrow F^n(i,\bar{x}))]$$

with this definition $Sat_\phi(A)$ is a Σ^1_n formula if ϕ is in Σ_{n+2}.

Let now $E \subset G$ be existentially universal and let ϕ be a Σ_3 formula with parameters $\bar{a}$ in E which is satisfied in G. Then $\mathcal{N}_G = \mathcal{N}_o \models Sat_\phi(A_{\bar{a}})$ and this is a Σ^1_1 formula. Thus $\mathcal{N}_E \models Sat_\phi(A_{\bar{a}})$ and $E \models \phi(\bar{a})$.

Let $Th(\beta_n)$ be the theory of the β_n models. Any countable β-model is induced by some regular model.

12.8 Corollary: If $\phi \in L^2$ then $\phi \in Th(\beta_n)$ iff $\bar{\phi} \in Th(\mathcal{E}_{n+1})$. If $\phi \in L$ then $\phi \in Th(\mathcal{E}_{n+1})$ iff $Sat_\phi \in Th(\beta_n)$.

Thus $Th(\beta_n)$ and $Th(\mathcal{E}_{n+1})$ are recursively equivalent.

In 12.6 we saw that every model in $\mathcal{E}_2$ is biregular. The conver does not hold:

12.9 Corollary: Every countable generic model has an elementary extension which is biregular but not in $\mathcal{E}_2$.

Proof: Let G be a countable generic model then $\mathfrak{N}_G \prec \mathfrak{N}_o$. By the main result of [67] $\mathfrak{N}_G$ has an elementary extension $\mathfrak{N}$ which is an ω-model but not a β_1 model. By 11.12 $\mathfrak{N} = \mathfrak{N}_E$ for some biregular model E and by 11.5 $G \prec E$. It follows now from 12.7 that $E \notin \mathcal{E}_2$.

Note that the model E above is an example of an existentially complete model of T^F which is not generic.

§3 The Analytic Hierarchy

Next we relate directly the hierarchy of the $\mathcal{E}_n$ classes to the analytic hierarchy. Given any standard Gödel enumeration all countable models in the language L can be interpreted as sets in $\mathfrak{N}_o$. Under such an enumeration the predicate $\phi(X)$ stating that X is a (countable) existentially complete model can be shown to be hyper-arithmetical $(\Pi^1_1 \cap \Sigma^1_1)$. More generally the predicate $\phi_{n+1}(X)$ which states that $X \in \mathcal{E}_{n+1}$ is Π^1_n (see 7.4). As $Th(\mathcal{E}_n)$ is the theory of the countable models in $\mathcal{E}_n$ we get that $Th(\mathcal{E}_n)$ is a Π^1_n set. We shall prove that $Th(\mathcal{E}_n)$ is a complete Π^1_n set - any Π^1_n set is recursively reducible to $Th(\mathcal{E}_n)$. For $n \geq 2$ this is almost immediate from 12.7: Given a Π^1_n set A described by the formu $\phi(x)$ ($x \in A$ iff $\mathfrak{N}_o \models \phi(x)$) then for $n \in A$ $T^F \vdash \overline{\phi(n)}$ (see 11.1) and for $n \notin A$ $T^F \vdash \sim\overline{\phi(n)}$. We shall show that the recursive correspondence between n and $\overline{\phi(n)}$ is such a reduction: $T(\mathcal{E}_n) \vdash \overline{\phi(n}$ iff $n \in A$. Clearly if $n \notin A$ then $T(\mathcal{E}_n) \nvdash \overline{\phi(n)}$ as $T(\mathcal{E}_n) \subseteq T^F$. On the other hand if $n \in A$ then $\mathfrak{N}_o \models \phi(n)$ where $\phi(n)$ is a Π^1_n sentence. It remains to show that for every $E \in \mathcal{E}_n$ $E \models \overline{\phi(n)}$ or equivalently $\mathfrak{N}_E \models \phi(n)$. But this is immediate from the fact that

$\mathcal{N}_E$ is a β_{n-1} model.

For $n = 1$ we need the following observation

12.10 Lemma: If $E, E' \in \mathcal{E}$ and if ϕ is a sentence with parameters of E that has only quantifiers relativited to $I(x)$ in front of a bounded formula then $E \models \phi$ iff $E' \models \phi$.

Proof: For bounded formulas this holds by 8.7. We proceed by induction on the number of relativited quantifiers. Assume $\psi = \exists x(I(x) \wedge \psi(x))$. If either of the models satisfies ϕ then it satisfies $\psi(n)$ for some n. By induction assumption the other one satisfies $\psi(n)$ and therefore ϕ. Taking the negations the same holds if $\phi = \forall x(I(x) \rightarrow \psi(x))$

It now follows that if $\phi(x)$ is a Π^1_1 formula and $\mathcal{N}_o \models \phi(n)$ then $E \models \overline{\phi(n)}$ whenever $E \in \mathcal{E}$. As above we get $Th(\mathcal{E}) \models \overline{\phi(n)}$ iff $\mathcal{N}_o \models \phi(n)$. Thus we have

12.11 Theorem: $Th(\mathcal{E}_n)$ is a complete Π^1_n set for $n = 1,2\ldots$.

The last result for $n = 1$ is quite surprising. $T_{\mathcal{E}}$ is a complete Π^1_1 set and any extension of $T_{\mathcal{E}}$ which is included in T^F is not Σ^1_1. On the other hand $T_{\mathcal{E}} \cup \{\forall x\ I(x)\}$ is $\mathcal{T}h(N)$ (by 8.26) which is known to be hyperarithmetical.

The $\mathcal{H}_n$ hierarchy is defined in chapter 4 using model theoretic methods. It decreases faster than the $\mathcal{E}_n$ hierarchy. Still using the same method as for the $\mathcal{E}_n$ hierarchy we can show

12.12 Theorem:

i) If $E \in \mathcal{H}_{n+1}$ then $\mathcal{N}_E$ is a β_{2n} model.

ii) $Th(\mathcal{H}_n)$ is a complete Π^1_{2n-1} set.

Note that we do not have the converse of (i) and it seems unlikely that E is necessarily in $\mathcal{H}_{n+1}$ (or even $\mathcal{H}_2$) whenever $\mathcal{N}_E$ is in β_{2n}.

12.13 In particular, as T^F is not analytical the different approximating chains for $\mathcal{J}$ are strictly decreasing.

12.14 We summarize the relations among the different classes of existentially complete models that were discussed in chapters 8-12.

i) $\mathcal{E}$ strictly includes the class of regular models, and the class of simple models. We were unable to provide an example of a model which is neither simple nor regular.

ii) The class of regular models includes the class of biregular models. We do not know if the inclusion is proper.

iii) The class of biregular models strictly includes $\mathcal{E}_2$.

iv) The sequences $\{\mathcal{E}_n\}$ and $\{\mathcal{H}_n\}$ are strictly decreasing (and the intersection of each is $\mathcal{G}$).

v) The class of generic models strictly includes the class of existentially universal models whose members are all uncountab

CHAPTER 13

APPLICATIONS TO COMPLETE EXTENSIONS OF PEANO'S ARITHMETIC

Let P be the set of axioms for Peano's arithmetic and T an arbitrary complete extension of P. Let $\sum = \sum_T$ be the class of models of $T_\forall$. It is easy to see that $\sum_T \supset \sum_{T(N)}$ and it follows from Matijasevic's theorem that the inclusion is sometimes proper. We shall see that for every such theory T there is a minimal existentially complete model N_T which has some of the properties that N has.

Replacing $T(N)$ by T in the definitions of chapter 8 we get the notions of $\sum_1$ formulas and recursive or partial recursive functions. Of course these definitions depend on T.

The analogues to the basic theorems of chapter 8 are the following

13.1 i) Every $\sum_1$ formula is equivalent in T to an existential formula.

ii) There is an $n+1$ place existential formula $F^n(y,\bar{x})$ which enumerates all existential formulas.

iii) There is a single formula $I(x)$ which defines N in any existentially complete model for any complete theory $P \subset T$.

The proof of i) is outlined in [39] and that of ii) and iii) in [63].

Thus theorems 8.10, 8.13 and 8.18 hold and we conclude that every submodel of an existentially complete model has an existential closure which coincides with the closure under partial recursive functions. We want to show that the existential closure of N is independent of the choice of the existentially complete extension E. As parameters in N have names in the language we conclude that the existential closure of N in E is $N_T(E) = \{a \in E \mid$ for some existential formula $e(x)$ $T \vdash \exists! x e(x)$ and $E \models e(a)\}$.

Let $M \models T_\forall \cup T_\exists$ be a model. Then $T \vdash \exists! x e(x)$ iff $M \models \exists! x e(x)$. Assume further that $E \cup M \subset M'$ and $M' \models T$ then the unique solution of each such formula e in M must coincide with the solution in E. Thus $N_T(E) \subset M$. As T is complete it has the joint embedding property.so that any two models in Σ can be embedded in a model of T. Hence $N_T(E)$ - is independent of the choice of E as long as $E \models T_\forall \cup T_\exists$ and we can denote it simply by N_T.

Finally if M is isomorphic to N_T and $M \cup N_T \subset M'$ then as M itself is existentially complete and therefore a model of $T_\forall \cup T_\exists$ we get from the remark above that $M = N_T$.

Thus we have:

13.2 <u>Theorem:</u> If $T \supset P$ is a complete theory then there is an existentially complete model N_T which can be embedded in a unique way in any model of $T_\forall \cup T_\exists$. Moreover, if $M \models T_\forall \cup T_\exists$ then

$$N_T = \{a \in M \mid T \vdash \exists! x e(x),\ M \models e(a),\ e \text{ is an existential formula}\}.$$

The introduction of the minimal models sheds some light on the subject of finite forcing in arithmetic. In [6] it was shown that N is the only finitely generic model for T(N). The situation is similar with other complete extensions of Peano's Arithmetic. To show this (13.4) we need the following lemma:

13.3 <u>Lemma:</u> If $M \equiv N_T$ and if M is existentially complete then $M = N_T$.

<u>Proof:</u> As $M \equiv N_T$, M is a model of $T_{\forall\exists}$ so that $N_T \subset M$. Using the formula I(x) which describes the natural numbers in N_T and the enumeration F^1 of existential formulas we have

$$N_T \models \forall x \exists z [I(z) \wedge \exists! y F^1(z,y) \wedge F^1(z,x)].$$

This sentence holds also in M. If M is also existentially complete I(x) defines exactly the set N in M so that every element of M is the unique solution of an existential formula without parameters. Thus $M = N_T$.

13.4 <u>Theorem:</u> If T is a complete extension of P then N_T is the unique finitely generic model for T.

<u>Proof:</u> Let G be any finite generic model. $G \models T_{\forall\exists}$ so that $N_T \subset G$ by 13.2, and N_T is existentially complete in G. Therefore N_T is finitely generic, and $N_T \equiv G$ as T has the joint embeddin

property. But G is also existentially complete and therefore by lemma 13.3 $G = N_T$.

PART THREE

DIVISION RINGS

A <u>division ring</u> or <u>skew field</u> is a ring in which each nonzero element has a multiplicative inverse. A division ring whose multiplication is commutative is a (commutative) field. The most familar example of a division ring with noncommutative multiplication is the ring of real quaternations. This ring is formed by the adjunction to the real numbers of elements i and j with the requirements that $i^2 = j^2 = -1$, $ij = -ji$ and i and j commute with all real numbers. A cyclic division ring, such as the ring of real quaternions, can be constructed from a finite dimensional, cyclic extension of a field whenever the norm mapping is not surjective. Another familar example of a division ring is the ring of endomorphisms of a faithful, irreducible module. Such division rings occur in the Jacobson density theorem and the Wedderburn-Artin theorems. A third example of a division ring is the coordinate ring of a projective geometry satisfying Desargues' axiom.

Since the concepts of existential completeness, infinite genericity, etc., were motivated by the example of algebraically closed fields, the investigation of these concepts for division rings is a natural undertaking. One must begin by specifying a language and a theory. Let $\mathcal{L}$ be the first order language with constant symbols 0 and 1 and function symbols $+$ and $\cdot$. As usual, $+(x,y)$ is denoted by $x + y$ and $\cdot(x,y)$ is denoted by xy. Let T be the following collection of axioms for division rings:

$$\sim(0 = 1)$$

$$\forall x \forall y \forall z \ ((x + y) + z = x + (y + z))$$

$$\forall x \forall y \ (x + y = y + x)$$

$$\forall x \ (x + 0 = x)$$

$$\forall x \exists y \ (x + y = 0)$$

$$\forall x \forall y \forall z \ ((xy)z = x(yz))$$

$$\forall x \ (x \cdot 1 = x \wedge 1 \cdot x = x)$$

$$\forall x \ (x = 0 \vee \exists y \ (xy = 1 \wedge yx = 1)$$

$$\forall x \forall y \forall z\ (x(y + z) = xy + xz \wedge (y + z)x = yx + zx).$$

The <u>characteristic</u> of a division ring is the least positive integer n for which the sum of 1 with itself n times equals 0, if such an n exists, and is 0 otherwise. The characteristic of a division ring is unique and is either a prime or zero. Let T_p for a prime p consist of the axioms of T and an axiom $1 + 1 + \ldots + 1 =$ where there are p occurrences of 1 on the left. Let T_0 consist of the axioms of T and axioms $\sim(1 + 1 + \ldots + 1 = 0)$, where there are n occurrences of 1 on the left, for all positive integers n. The models of T_p are the division rings of characteristic p and the models of T_0 are the division rings of characteristic 0.

The <u>center</u> of a division ring is the set of elements of the ring which commute with all elements of the ring. The center of a division ring is a division ring, and its multiplication is commutative by definition; hence, the center is a field. The axioms of T imply that 0 and 1 are always in the center. Consequently, the center of a division ring contains either the rational numbers or the field with p elements, according to whether the characteristic is 0 or p.

That the center of a division ring should include a particular field is a frequently encountered requirement in algebra (see (2), (3), (21) - (28), (48)). The class of division rings whose centers include a specified field k is axiomatized as follows. Add names for all elements of k to $\mathcal{L}$ to form a new language $\mathcal{L}(k)$. Let T_k consist of the axioms of T, the diagram of k in the language $\mathcal{L}(k)$, and axioms $\forall x(xa = ax)$ as a ranges over all elements of k. The models of T_k are precisely the division rings whose centers include k. A model of T_k is called a <u>division algebra over k</u>. Note that the models of T_0 and T_p are the same as those of $T_{\mathbb{Q}}$ and T_{F_p}, respectively, where $\mathbb{Q}$ denotes the rational numbers and F_p denotes the field with p elements.

Algebraic properties of existentially complete division algebras are discussed in Chapter 14. Some of the topics are subfields, maximal subfields, centralizers, and endomorphisms.

The Nullstellensatz for existentially complete division algebras is presented in Chapter 15. Noncommutative polynomials and a suitable radical are defined. Then the Nullstellensatz for finitely generated polynomial ideals is proven. A restricted version of the Nullstellensatz is proven for infinitely generated polynomial ideals.

The model theoretical questions arising from Part One are answered in Chapter 16. The two principal results are that the class of infinitely generic division algebras is disjoint from the class of finitely generic division algebras and that there are $2^{\aleph_0}$ non-elementarily equivalent existentially complete division algebras.

Many of the results in Chapter 14 and Chapter 16 are also true when the phrase "division algebra" is replaced by "group". Specific details are described in footnotes.

Most of the results on the theories T, T_0, and T_p appear also in W. Wheeler (107).

The following terminology will be used: a member of $\mathcal{E}_T$ will be called an existentially complete division ring; a member of $\mathcal{E}_{T_0}$ or $\mathcal{E}_{T_p}$ will be called an existentially complete division ring of characteristic 0 or characteristic p, respectively; a member of $\mathcal{E}_{T_k}$ will be called an existentially complete division algebra over k; and so forth for the other classes. If D is a division algebra over K and S is a subset of D, then $\langle S\rangle$ will denote the division subalgebra over k generated by S.

Three methods of constructing division algebras will be utilized in the succeeding chapters: skew polynomial rings, skew power series rings, and amalgamation of division algebras. These constructions are described briefly in the following paragraphs.

The first method is that of skew polynomial rings (Ore (72)).

Suppose D is a division algebra over k. Let $f : D \to D$ be an endomorphism of D, and let $\delta : D \to D$ be an f-derivation, that is, a map satisfying $\delta(a + b) = \delta(a) + \delta(b)$ and $\delta(ab) = \delta(a)\cdot f(b) + a\cdot\delta(b$ for all elements a and b of D. Let x be an indeterminate, and consider the set of polynomials in x with coefficients from D on the right. The general polynomial has the form $a_0 + xa_1 + \ldots + x^n a_n$. Addition is defined as usual. Multiplication of two polynomials is defined in the usual way using the commutation rule $ax = x\cdot f(a) + \delta(a)$. The collection of right polynomials with these operations of addition and multiplication is a right Ore domain, denoted by $D[x; f, \delta]$. As a right Ore domain, $D[x; f, \delta]$ has a unique right quotient ring, which is a division ring. This division ring is denoted by $D(x; f, \delta)$.

Skew power series may be defined in an analogous manner when $\delta = 0$ Let $f : D \to D$ be an endomorphism of a division algebra D. Let x be an indeterminate. Consider the collection of formal Laurent series $\sum_{i=n}^{\infty} x^i a_i$, where n is an integer and the a_i's are elements of D. Addition is defined as usual. Multiplication is defined in the usual manner using the commutation rule $ax = x\cdot f(a)$. The collection of formal Laurent series with these operations is a division algebra, denoted by $D((x; f))$.

The third method is amalgamation of division algebras. Suppose that D_1 and D_2 are division algebras over a field k and that E_1 and E_2 are division subalgebras of D_1 and D_2, respectively, such that E_1 and E_2 include k and are isomorphic. P. M. Cohn (21, 27) has shown that the diagram

$$\begin{array}{ccc} D_1 & & D_2 \\ \cup\!\text{I} & & \cup\!\text{I} \\ E_1 & \cong & E_2 \end{array}$$

can be completed to

$$\begin{array}{ccccc} & & D_3 & & \\ & \nearrow & & \nwarrow & \\ D_1 & & & & D_2 \\ \cup & & & & \cup \\ E_1 & & \cong & & E_2 \end{array}$$

in the category of division algebras over k and that there is a universal choice for D_3 as the initial object in a certain category. This universal division algebra over k will be called the amalgamation of D_1 and D_2 and will be denoted by $D_1 \underset{E_1}{*} D_2$. An important property of the amalgamation is that in $D_1 \underset{E_1}{*} D_2$, $D_1 \cap D_2 = E_1$.

CHAPTER 14

EXISTENTIALLY COMPLETE DIVISION RINGS

The arithmetic of existentially complete division rings is quite intricate. The arithmetic is so complex as to preclude the construction of these rings through successive adjunctions of algebraic elements, in contrast to the case of algebraically closed fields. The intricacy of the arithmetic is reflected in the actions of the inner automorphisms of the ring. These actions can be used to formulate first order definitions of concepts which usually require a second order language or an infinitary language. Examples of such concepts are the division subalgebra generated by a finite subset, the property of being transcendental over the center, and the set of nonnegative powers of a transcendental element (other examples appear in M. Boffa and P. van Praag (11), A. Macintyre (58), and W. Wheeler (107)). In turn, these first order definitions reduce the study of structural properties to the study of arithmetical properties.

This chapter consists of five sections. The basic lemmas on inner automorphisms are presented in the first section, where they are used to show that the center of an existentially complete division algebra over a field k is just k. In the second section, the actions of inner automorphisms are used to give an arithmetical description of elements which are transcendental over k. The existence of certain subfields of existentially complete division algebras is established also. Centralizers and maximal subfields of countable, existentially complete division algebras are discussed in the third section. In particular, it is shown that each countable, nonfinitely generated extension field of the center occurs as a maximal subfield. In the fourth section, the existence of a proper endomorphism is proven. This endomorphism is used to construct extensions with no new finitely generated division subalgebras. Finally, in the fifth section, inner automorphisms are used to construct $2^{\aleph_0}$ nonisomorphic division algebras generated by two elements.

Let k be a commutative field. All division rings will be division algebras over k, unless otherwise specified. All division subalgebras are required to include k. An element of a division algebra will be called transcendental if it is transcendental over k.

§ 1 Inner Automorphisms

The utilization of inner automorphisms depends upon the following result of P. M. Cohn (21)

Theorem 14.1. If D is a division algebra over k, E and E' are division subalgebras of D, and $f : E \to E'$ is an isomorphism of E onto E' over k, then there is a division algebra D' over k extending D and an element d' of D' such that $(d')^{-1}ad' = f(a)$ for all elements a of E.

Proof. Let $k(x)$ and $k(y)$ be the fields of rational functions over k in indeterminates x and y, respectively. Let H_1 and H_2 be the amalgamations $D \underset{k}{*} k(x)$ and $D \underset{k}{*} k(y)$, respectively. Let D' be the amalgamation of H_1 and H_2 with amalgam

$$\langle D, x^{-1}Ex \rangle \cong \langle D, y^{-1}E'y \rangle ,$$

where the isomorphism is given by $d \leftrightarrow d$, $x^{-1}\ ax \leftrightarrow y^{-1}\ f(a)y$. In D', $(xy^{-1})^{-1}\ a(xy^{-1}) = f(a)$ for all elements a of E.

Suppose that the division subalgebras E and E' in the preceding theorem are finitely generated over k, say $E = \langle a_0, \ldots, a_n \rangle$, $E' = \langle b_0, \ldots, b_n \rangle$, and that f is given by $f(a_i) = b_i$. Then the element d' in the extension D' satisfies $(d')^{-1}\ a_i\ d' = b_i$ for $i = 0, \ldots, n$. Therefore, D' satisfies the sentence

$$\exists v\ (a_0v = vb_0 \wedge \ldots \wedge a_nv = vb_n \wedge \sim(v = 0)).$$

If D is existentially complete, then D must satisfy this sentence also. This proves the following lemma.

Lemma 14.2. If D is an existentially complete division algebra over k and $E = \langle a_0, \ldots, a_n \rangle$ and $E' = \langle b_0, \ldots, b_n \rangle$ are finitely generated division subalgebras, then $a_i \to b_i$ for $i = 0, \ldots, n$ determines an isomorphism of E onto E' if and only if there is an element d of D such that $d^{-1}a_id = b_i$ for $i = 0, \ldots, n$.

A division algebra over k with the property that any isomorphism of two finitely generated division subalgebras can be extended to an automorphism of the entire division algebra is said to be finitely homogeneous.

The centralizer of a subset S in a division algebra D, denoted by $C_D(S)$ is the set of elements of D which commute with all elements of S.

Lemma 14.3. Let D be an existentially complete division algebra over k. An element b of D is in the division subalgebra of D generated by elements $a_0, \ldots, a_n$ of D if and only if the centralizer of b in D includes the centralizer of $\{a_0, \ldots, a_n\}$ in D.

Proof. Necessity is clear. In order to prove sufficiency, suppose b is not in the division subalgebra $\langle a_0, \ldots, a_n \rangle$. Let E' be a division algebra over k isomorphic to $\langle a_0, \ldots, a_n, b \rangle$ with generators $a_0', \ldots, a_n', b'$, and assume that $D \cap E' \subseteq k$. Amalgamate D and E' identifying $\langle a_0, \ldots, a_n \rangle$ with $\langle a_0', \ldots, a_n' \rangle$ in the natural way, and let D' be an existentially complete division algebra extending this amalgamation. According to Lemma 14.2, there is an element d' of D' such that $(d')^{-1}bd' = b'$ and $(d')^{-1}a_id' = a_i'$ for $i = 0, \ldots, n$. Since b is not in $\langle a_0, \ldots, a_n \rangle$, b and b'

are distinct in D'. Consequently, D' satisfies the sentence

$$\exists v(\sim(v = 0) \wedge a_0 v = va_0 \wedge \ldots \wedge a_n v = va_n \wedge \sim(bv = vb)).$$

Since D is existentially complete, D satisfies this sentence also. Thus, the centralizer of b in D does not include the centralizer of $\{a_0, \ldots, a_n\}$ in D.

Corollary 14.4. If E is a finitely generated division subalgebra of an existentially complete division algebra D over k, then $C_D(C_D(E)) = E$.

Corollary 14.5. The center of an existentially complete division algebra over k is just k. In particular, the center of an existentially complete division ring of characteristic of 0 or p is just the rational numbers or the field with p elements, respectively.

Proof. Suppose D is an existentially complete division algebra over k. The subfield k of D is the division subalgebra of D generated by 1. If an element b of D does not lie in k, then, according to Lemma 14.3, some element of D does not commute with b. Therefore, the center of D is k.

Proposition 14.6. If E is a finitely generated division subalgebra of an existentially complete division algebra D over k, then $C_D(E)$ is not finitely generated over k.

Proof. Let s and t be indeterminates. Let $D' = D(s; 1_D, 0_D)$ where $1_D : D \to D$ is the identity mapping and $0 : D \to D$ is zero mapping, that is, $0_D(a) = 0$ for all a in D. Let $D'' = D'(t; f, 0_{D'})$ where $f : D' \to D'$ is defined by $f(d) = d$ for all elements d of D and $f(s) = s^2$.

Assume that E is generated by $e_0, \ldots, e_n$. Suppose $d_0, \ldots, d_m$ are elements of $C_D(E)$. The element s of D'' is in the centralizer of E in D'', and the element t of D'' commutes with the elements

$d_0, \ldots, d_m$ but not with s. Therefore, D'' satisfies the sentence

$$\exists x \exists y\ (\sim(x = 0) \wedge \sim(y = 0) \wedge e_0x = xe_0 \wedge \ldots \wedge e_nx = xe_n \wedge d_0y = yd_0 \wedge \ldots \wedge d_my = yd_m \wedge \sim(xy = yx)).$$

The division algebra D must satisfy this sentence also, since D is existentially complete. In other words, there is an element of $C_D(E)$ whose centralizer in D does not include $C_D(\{d_0, \ldots, d_n\})$. Hence, according to Lemma 14.3, this element of $C_D(E)$ is not in the division subalgebra generated by $d_0, \ldots, d_n$.

Corollary 14.7. An existentially complete division algebra over k cannot be finitely generated.

Proof. Apply the preceding proposition to $C_D(k) = C_D(\langle 1 \rangle) = D$.

Proposition 14.8. The theory of existentially complete division rings of characteristic zero is not an arithmetical set.

Proof. The center of an existentially complete division ring of characteristic 0 is the rational numbers. Consequently, an interpretation of the theory of the rational numbers within $\mathcal{T}h(\mathcal{E}_{T_0})$ is obtained by relativizing each sentence in the language of the rational numbers to the center of the division algebra. That the theory of the natural numbers is interpretable within the theory of the rational numbers is well-known (J. Robinson (86)). Hence, the theory of the natural numbers is interpretable within the theory of existentially complete division rings of characteristic 0.

§ 2 Transcendental Elements and Subfields

The discussion of the commutative subfields of an existentially complete division algebra depends upon the fact that the property of being transcendental over a finitely generated subfield is arithmeticall definable in an existentially complete division algebra. One may recall that the property of being transcendental is not an elementary property in the theory of algebraically closed fields. The disparity is due to the action of the inner automorphisms in existentially complete division algebras.

Define formulas $\mathrm{Transc}_n(v_1, \ldots, v_n)$ for all positive integers n by induction as follows:

$$\begin{aligned}\mathrm{Transc}_1(v_1) \text{ is } & \sim(v_1 = 0) \wedge \exists y \exists z(\sim(y = 0) \wedge \sim(z = 0) \\ & \wedge v_1 y = y v_1{}^2 \wedge v_1{}^2 z = z v_1{}^2 \\ & \wedge \sim(v_1 z = z v_1)) \quad ;\end{aligned}$$

$$\begin{aligned}\mathrm{Transc}_n(v_1, \ldots, v_n) \text{ is } & \sim(v_n = 0) \wedge \mathrm{Transc}_{n-1}(v_1, \ldots, v_{n-1}) \\ & (v_1 v_n = v_n v_1 \wedge \ldots \wedge v_{n-1} v_n = v_n v_{n-1}) \\ & \wedge \exists y \exists z(\sim(y = 0) \wedge \sim(z = 0) \\ & \wedge v_1 y = y v_1 \wedge \ldots \wedge v_{n-1} y = y v_{n-1} \wedge v_n y = y v_n{}^2 \\ & \wedge v_1 z = z v_1 \wedge \ldots \wedge v_{n-1} z = z v_{n-1} \\ & \wedge v_n{}^2 z = z v_n{}^2 \wedge \sim(v_n z = z v_n)).\end{aligned}$$

The formula $\mathrm{Transc}_1(v_1)$ will be denoted by $\mathrm{Transc}(v)$ also.

Lemma 14.9. (i) If D is a division algebra over k and $a_1, \ldots, a_n$ are elements of D for which D satisfies $\mathrm{Transc}_n(v_1, \ldots, v_n)$, then $a_1, \ldots, a_n$ generate a commutative subfield of D of transcendence degree n over the center of D.

(ii) If D is an existentially complete division algebra over k and $a_1, \ldots, a_n$ are elements of D which commute and are algebraically independent over the center of D, then D satisfies $\mathrm{Transc}_n(a_1, \ldots, a_n)$.

Proof. (i) Assume $\mathrm{Transc}(a)$ holds in D. Then neither a nor a^2 is in the center of D. Since there is a nonzero element d of D such that $ad = da^2$, i.e., $d^{-1}ad = a^2$, a and a^2 generate isomorphic subfields over the center C of D. Therefore,

$$[C(a^2) : C] = [C(a) : C] = [C(a) : C(a^2)] \cdot [C(a^2) : C] .$$

Since there is an element of D which commutes with a^2 but not with a, a is not in $C(a^2)$; so $[C(a) : C(a^2)] = 2$. Then $[C(a^2) : C] = 2 \cdot [C(a^2) : C]$, so the dimension of $C(a^2)$ over C must be infinite. Hence, a is transcendental over C.

Assume now that part (i) holds for some $n \geq 1$. Suppose D satisfies $\mathrm{Transc}_{n+1}(a_1,\ldots, a_n, a_{n+1})$. The elements $a_1, \ldots, a_n$ commute and are algebraically independent over C by the induction hypothesis. Repeating the preceding argument with C replaced by $C(a_1,\ldots, a_n)$ shows that $a_1, \ldots, a_n, a_{n+1}$ commute and are algebraicall independent over C.

(ii) Suppose a is transcendental over the center of D. The center of D, according to Corollary 14.5, is just the field k. The subfields k(a) and $k(a^2)$ are isomorphic, since a is transcendental over k. According to Lemma 14.2, there is an element d_1 of D such that $d_1 \neq 0$ and $d_1^{-1}ad_1 = a^2$. Since a is not in the field $k(a^2)$, Lemma 14.3 asserts that there is an element d_2 of D such that $d_2 \neq 0$, $a^2d_2 = d_2a^2$, and $ad_2 \neq d_2a$. Thus, D satisfies Transc(a).

The case of $n > 1$ follows by induction using the same argument with k replaced by $k(a_1,\ldots, a_{n-1})$.

Theorem 14.10. Each countably generated extension field of k is included in every existentially complete division algebra over k.

Proof. It suffices to prove the result for extension fields with transcendence degree $\aleph_0$ over k. Assume L is such an extension field of k. Let $\{a_i : 1 \leq i < \omega\}$ be a transcendence basis for L over k and let $\{b_i : 1 \leq i < \omega\}$ be a sequence of elements of L such that b_n is algebraic over $k(a_1,\ldots, a_n, b_1,\ldots, b_{n-1})$ and L is generated by $\{a_i, b_i : 1 \leq i < \omega\}$. For each $n \geq 1$, let $p_n(x_1,\ldots, x_n, y_1,\ldots, y_n)$ be the rational function over k for which $p_n(a_1,\ldots, a_n, b_1,\ldots, b_{n-1}, y_n)$ is the monic irreducible polynomial of b_n over $k(a_1,\ldots, a_n, b_1,\ldots, b_{n-1})$. Define formulas ϕ_n for positive integers n inductively as follows:

ϕ_1 is $\mathrm{Transc}_1(x_1) \wedge x_1y_1 = y_1x_1 \wedge p_1(x_1,y_1) = 0$;

ϕ_{n+1} is $\phi_n \wedge \mathrm{Transc}_{n+1}(x_1,\ldots, x_{n+1})$

$\wedge \bigwedge_{i=1}^{n} (x_{n+1}y_i = y_ix_{n+1} \wedge x_iy_{n+1} = y_{n+1}x_i \wedge y_iy_{n+1} = y_{n+1}y_i)$

$\wedge x_{n+1}y_{n+1} = y_{n+1}x_{n+1}$

$\wedge p_{n+1}(x_1,\ldots, x_{n+1}, y_1,\ldots, y_{n+1}) = 0$.

Let D' be an existentially complete division algebra over k which extends the amalgamation $D \underset{k}{*} L$. The division algebra D' satisfies $\exists x_1 \ldots \exists x_n \exists y_1 \ldots \exists y_n \phi_n$ for each positive integer n. Therefore, the division algebra D must satisfy all these sentences also, since D is existentially complete.

Suppose elements $c_1, \ldots, c_n, d_1, \ldots, d_n$ of D have been chosen for which D satisfies $\phi_n(c_1,\ldots, c_n, d_1,\ldots, d_n)$. Since D satisfies $\exists x_1 \ldots \exists x_{n+1} \exists y_1 \ldots \exists y_{n+1} \phi_{n+1}$, there are elements $c_1', \ldots, c_{n+1}', d_1', \ldots, d_{n+1}'$ for which D satisfies $\phi_{n+1}(c_1',\ldots, c_{n+1}', d_1',\ldots, d_{n+1}')$. Then $\langle c_1,\ldots, c_n, d_1,\ldots, d_n\rangle$ and $\langle c_1',\ldots, c_n', d_1',\ldots, d_n'\rangle$ are isomorphic, so there is an element e of D for which $e^{-1}c_i'e = c_i$, $e^{-1}d_i'e = d_i$ for $i = 1, \ldots, n$. Let $c_{n+1} = e^{-1}c_{n+1}'e$, $d_{n+1} = e^{-1}d_{n+1}'e$. Then D satisfies $\phi_{n+1}(c_1,\ldots, c_{n+1}, d_1,\ldots, d_{n+1})$, so the field generated by $c_1, \ldots, c_{n+1}, d_1, \ldots, d_{n+1}$ is isomorphic to the subfield of L generated by $a_1, \ldots, a_{n+1}, b_1, \ldots, b_{n+1}$.

This procedure yields a set $\{c_i, d_i : 1 \leq i < \omega\}$ of elements of D which generates a subfield isomorphic to L over k.

<u>Corollary 14.11</u>. Each existentially complete division algebra over k includes a commutative subfield of infinite transcendence degree over k.

<u>Corollary 14.12</u>. If k is countable, then each existentially complete division algebra over k includes an algebraically closed field of infinite transcendence degree over k.

Corollary 14.5 and Corollary 14.12 will be used to prove the following result on the model theory of division algebras.

Theorem 14.13. The theory T_k of division algebras over k does not have a model-companion.

Proof. It suffices to show that the class of existentially comple division algebras over k is not closed under ultrapowers. Let D be an existentially complete division algebra over k. Let $\mathcal{U}$ be a nonprincipal ultrafilter on ω, and consider the ultrapower $D^{\omega}/\mathcal{U}$. First, suppose k is infinite. Choose a sequence $a_0, a_1, a_2, \ldots$ of elements of k such that $a_i \neq a_j$ for $i \neq j$. The element (a_i) of $D^{\omega}/\mathcal{U}$ is in the center of $D^{\omega}/\mathcal{U}$ but is not in k. Therefore, D cannot be an existentially complete division algebra over k, because its center is strictly larger than k. On the other hand, suppose k is finite. Let $\{p_n(x) : n < \omega\}$ be an enumeration of all monic irreducible polynomials over k such that each polynomial is listed only once. Each polynomial in this enumeration has a zero in D, according to Corollary 14.12. For each n, let a_n be a zero of $p_n(x)$ in D and let b_n be a zero of $p_{n+1}(x)$ in D. Both elements (a_i) and (b_i) of the ultrapower $D^{\omega}/\mathcal{U}$ are transcendental over the center of $D^{\omega}/\mathcal{U}$. However, there is no inner automorphism mapping (a_i) to (b_i), because a_n and b_n generate nonisomorphic subfields over k in each factor. Therefore, $D^{\omega}/\mathcal{U}$ is not an existentially complete division algebra over k.

The preceding proof is essentially the same as the proof of P. Eklof and G. Sabbagh (34) that the theory of groups has no model-companion. Not surprisingly, then, the preceding theorem for the cases of T, T_0, and T_p was proven independently by G. Sabbagh (91), A. Macintyre (58), M. Boffa and P. van Praag (11), and W. Wheeler (107)

§ 3 Centralizers and Maximal Subfields

A subfield of a division algebra is said to maximal if it is not included in any strictly larger subfield. Although Theorem 14.10 and Corollaries 14.11 and 14.12 assert the existence of certain subfields, these results do not indicate whether the subfields are maximal. After the appearence of an earlier version of Corollary 14.12 (W. Wheeler (107)), M. Boffa and P. van Praag(12) proved that each countable, infinitely generic division ring of characteristic 0 or p included as maximal subfields an algebraically closed field of infinite transcendence degree over the center, an algebraically closed field which was algebraic over the center, and a purely transcendental field of infinite transcendence degree over the center. These results will be subsumed in a characterization of all maximal subfields of countable, existentially complete division algebras over k.

If D_1 and D_2 are division algebras over k and D_2 extends D_1, then a copy $D_1' \subseteq D_2'$ of $D_1 \subseteq D_2$ is a pair of division algebras over k for which there is an isomorphism f (over k) of D_2 onto D_2' with $f(D_1) = D_1'$.

A subfield of a divison algebra is maximal if and only if it equals its centralizer. The following result on centralizers is therefore the crux of the characterization of maximal subfields.

Proposition 14.14. Assume that k is a countable field and that D is a countable, existentially complete division algebra over k. Suppose D_1 and D_2 are division subalgebras of D such that D_1 is included in D_2 but D_1 is not included in any finitely generated division subalgebra of D_2. Then there is a copy $D_1' \subseteq D_2'$ of $D_1 \subseteq D_2$ such that $C_D(D_1') = C_{D_2'}(D_1')$.

Proof. Let $\{d_n : n < \omega\}$ be an enumeration of $D - k$, $\{a_n : n < \omega\}$ be an enumeration of $D_2 - k$, and $\{b_n : n < \omega\}$ be an enumeration of $D_1 - k$. Construct a copy of $D_1 \subseteq D_2$ inductively as follows.

Step 0 : Let $a_0' = a_0$, $b_0' = b_0$.

Step n + 1 : Assume elements $a_0', \ldots, a_{r_n}'$ $b_0', \ldots, b_{r_n}'$ have been selected such that (i) there is an element c of D such that $c^{-1}a_ic = a_i'$ and $c^{-1}b_ic = b_i'$ for $i = 0, \ldots, r_n$, and (ii) for each $i < n$ (if $n \neq 0$), either d_i is in the division subalgebra generated by $a_0', \ldots, a_{r_n}'$, $b_0', \ldots, b_{r_n}'$ or d_i is not in $C_D(b_0', \ldots, b_{r_n}')$. If d_n is in the division subalgebra generated by

$a_0', \ldots, a_{r_n}'$, $c^{-1}(a_{r_n+1})c$, $b_0', \ldots, b_{r_n}'$, $c^{-1}(b_{r_n+1})c$ or is not in $C_D(\{b_0', \ldots, b_{r_n}', c^{-1}(b_{r_n+1})c\})$, then let $r_{n+1} = r_n + 1$, $a_{r_{n+1}}' = c^{-1}(a_{r_{n+1}})$ $b_{r_{n+1}}' = c^{-1}(b_{r_{n+1}})c$. Otherwise, let r_{n+1} be the least integer j greater than r_n for which b_j is not in the division subalgebra generated by $a_0, \ldots, a_{r_n}$, $b_0, \ldots, b_{r_n}$. Such an integer exists, because D_1 is not included in any finitely generated division subalgebra of D_2 by hypothesis. Amalgamate D with a copy D_2'' (disjoint from D) of D_2 identifying $a_i' \leftrightarrow a_i''$, $b_i' \leftrightarrow b_i''$ for $i = 0, \ldots, r_n$. The element $b_{r_{n+1}}''$ does not commute with d_n in this amalgamation (see P. M. Cohn (21, 27)). Let E be an existentially complete division algebra over k which includes this amalgamation. The division algebra E satisfies the sentence

$$\exists x \exists y(\sim(x = 0) \wedge \bigwedge_{i=1}^{r_n} (a_ix = xa_i' \wedge b_ix = xb_i') \wedge b_{r_{n+1}}x = xy \wedge \sim(d_ny = yd_n)).$$

The division algebra D must satisfy this sentence also, because D is existentially complete. Let c' be an element of D for which D satisfies

$$\exists y(\sim(c' = 0) \wedge \bigwedge_{i=1}^{r_n} (a_ic' = c'a_i' \wedge b_ic' = c'b_i') \wedge b_{r_{n+1}}c' = c'y \wedge \sim(d_ny = yd_n)).$$

Let $a_i' = (c')^{-1}a_ic'$, $b_i' = (c')^{-1}b_ic'$ for

$i = r_n + 1, \ldots, i = r_{n+1}$. This completes step $n + 1$.

Let D_2' be the division subalgebra generated by $\{a_n' : n < \omega\}$ and let D_1' be the division subalgebra generated by $\{b_n' : n < \omega\}$. Clearly, $D_1' \subseteq D_2'$ is a copy of $D_1 \subseteq D_2$. Suppose d is an element of D. Then $d = d_n$ for some nonnegative integer n. The construction above ensures that either d is in the division subalgebra generated by $a_0', \ldots, a_{r_{n+1}}', b_0', \ldots, b_{r_{n+1}}'$ or d does not commute with one of the elements $b_0', \ldots, b_{r_{n+1}}'$. Hence, $C_D(D_1') = C_{D_2'}(D')$.

Corollary 14.15. If K is an extension field of k and L is an extension field of K such that L is countable and K is not finitely generated over k, and D is a countable, existentially complete division algebra over k, then D includes a copy $K' \subseteq L'$ of $K \subseteq L$ such that $C_D(K') = C_{L'}(K') = L'$.

Proof. The division algebra D includes a copy of $K \subseteq L$, since D includes an algebraically closed field of infinite transcendence degree over k. The corollary follows by applying Proposition 14.14.

Theorem 14.16. The following are equivalent for a countable, extension field K of a countable field k:

(i) K occurs as a maximal subfield of an existentially complete division algebra over k;

(ii) K is not finitely generated over k;

(iii) K occurs as a maximal subfield of every countable, existentially complete division algebra over k.

Proof. (i) implies (ii). Suppose K is a maximal subfield of an existentially complete division algebra D over k. Then $C_D(K) = K$. Since a finitely generated division subalgebra of D cannot be its own centralizer (Proposition 14.6), K is not finitely generated over k.

(ii) implies (iii). Apply Corollary 14.15 with $K = L$.

(iii) implies (i). Obvious.

Corollary 14.17. If K is a countable, nonfinitely generated, extension field of a countable field k and K_0 is a subfield of K which is finitely generated over k, then each copy of K_0 in a countable, existentially complete division algebra over k is included in a maximal subfield which is isomorphic over K_0 to K.

Proof. Suppose K_0' is a copy of K_0 in an existentially complet division algebra D over k. There is a copy $K_0'' \subseteq K''$ in D of $K_0 \subseteq K$ for which K'' is a maximal subfield of D. Also, there is an automorphism f of D such that $f(K_0'') = K_0'$, since K_0 is finitely generated over k. Then $f(K'')$ is a maximal subfield of D, includes K_0', and is isomorphic over K_0' to K.

That each finitely generated division subalgebra E of an existentially complete division algebra D is its own double centraliz that is, $C_D(C_D(E)) = E$, was proved earlier. In fact, this is a special case of a more general result that some copy of each division subalgebra is its own double centralizer.

First, some remarks on notation are necessary. Suppose D is a division algebra, 1_D is the identity mapping of D onto itself, f is an endomorphism of D, and 0_D is the mapping of D onto $\{0\}$. The algebras $D[x; f, 0_D]$ and $D(x; f, 0_D)$ will be denoted by $D[x; f]$ and $D(x; f)$, respectively. The algebras $D[x; 1_D, 0_D]$ and $D(x; 1_D, 0_D)$ will be denoted by $D[x]$ and $D(x)$. The latter cases will also be described by saying that x is a commuting indeterminate.

Lemma 14.18. Let D_1 be a division algebra over k, and let $D_2 = (D_1(s))(t; f)$, where s and t are indeterminates, and $f : D_1(s) \to D_1(s)$ is defined by $f(d) = d$ for all d in D_1 and $f(s) = s^2$. The centralizer of $\{s, t\}$ in D_2 is D_1.

Proof. Adjoin s and all its roots $s^{1/n}$ for positive integers n to D_1 as commuting indeterminates to form a division algebra $\overline{D}_1$ over k. Consider the division algebra $E = \overline{D}_1((t; \overline{f}))$ of formal Laurent series, where $\overline{f}$ is defined by $\overline{f}(d) = d$ for all d in D_1 and $\overline{f}(s^r) = s^{2r}$ for any rational number r. Embed D_2 in E in the natural way.

The centralizer of $\{s, t\}$ in E is D_1. In order to verify this suppose that $\sum_{i=m}^{\infty} t^i R_i(s^{r_i})$, where m is an integer, r_i is a rational number, and $R_i(s^{r_i})$ is a rational function in the division algebra $\overline{D}_1$, commutes with both s and t. Then

$$\sum_{i=m}^{\infty} t^i R_i(s^{r_i})s = (\sum_{i=m}^{\infty} t^i R_i(s^{r_i}))s = s(\sum_{i=m}^{\infty} t^i R_i(s^{r_i}))$$

$$= \sum_{i=m}^{\infty} st^i R_i(s^{r_i}) = \sum_{i=m}^{\infty} t^i s^{2^i} R_i(s^{r_i}) = \sum_{i=m}^{\infty} t^i R_i(s^{r_i}) s^{2^i}, \text{ so } R_i(s^{r_i}) = 0$$

except possibly for $i = 0$. Suppose $R_0(s^{r_0}) \neq 0$. One may assume without loss of generality that $r_0 = 1$. The given Laurent series commutes with t by assumption, so $tR_0(s) = R_0(s)t = tR_0(s^2)$, that is, $R_0(s) = R_0(s^2)$. There are polynomials in $D_1[s]$ such that $p(s)/q(s) = R_0(s) = R_0(s^2) = p(s^2)/q(s^2)$. The degree of the polynomial $q(s)$ is either zero or greater than zero. If the degree of $q(s)$ is zero, then $q(s)$ is an element of D_1. Then $p(s) = p(s^2)$, so $p(s)$ is an element of D_1 also, so $R_0(s)$ is an element of D_1. Suppose, on the other hand, that the degree of $q(s)$ is greater than zero. Since $D_1[s]$ is a right Ore domain, there are nonzero polynomials $g(s)$ and $h(s)$ such that $q(s)g(s) = q(s^2)h(s)$ and $p(s)g(s) = p(s^2)h(s)$. These equalities imply that degree $q(s)$ = degree $g(s)$ - degree $h(s)$ = degree $p(s)$. Since s commutes with all elements of D_1, $D_1[s]$ is also a left Ore domain, so $p(s) = aq(s) + r(s)$ where a is an element of D_1, $r(s)$ is a polynomial in $D_1[s]$, and $r(s) = 0$ or the degree of $r(s)$ is less than the degree of $q(s)$. Then

$r(s)/q(s) = p(s)/q(s) - a$, and $r(s^2)/q(s^2) = t^{-1}(r(s)/q(s))t$ $= t^{-1}(p(s)/q(s) - a)t = t^{-1}(p(s)/q(s))t - t^{-1}at = p(s^2)/q(s^2) - a$ $= p(s)/q(s) - a = r(s)/q(s)$. As above, the equality $r(s^2)/q(s^2) = r(s)/q(s)$ implies that $r(s) = 0$ or degree $r(s) =$ degree $q(s)$. But either $r(s) = 0$ or degree $r(s) <$ degree $q(s)$, so $r(s) = 0$ and $p(s)/q(s)$ is an element of D_1. Thus, in either case, $R_0(s)$ is an element of D_1. Therefore, $D_1 \subseteq C_{D_2}(\{s, t\}) \subseteq C_E(\{s, t\}) \subseteq D_1$, so $D_1 = C_{D_2}(\{s, t\})$.

Let D_1 be a division algebra over k. Construct an ascending chain $D_1 \subseteq D_2 \subseteq D_3 \subseteq \ldots$ of division algebras over k by defining $D_{n+1} = (D_n(s_n))(t_n; f_n)$ where s_n and t_n are indeterminates and $f_n(d) = d$ for d in D_n, $f_n(s_n) = s_n^2$. Let $D_\omega = \bigcup_{0<n<\omega} D_n$.

Lemma 14.19. $C_{D_\omega}(\{s_i, t_i : i = 1, 2, \ldots\}) = D_1$.

Proof. Suppose a is an element of $C_{D_\omega}(\{s_i, t_i : i = 1, 2, \ldots$ Let n be the least positive integer for which a is in D_n. Suppose n is greater than one. Then a is in $C_{D_n}(\{s_{n-1}, t_{n-1}\})$, so, by the preceding lemma, a is in D_{n-1} contrary to the choice of n. Hence a is in D_1.

Conversely, each element of D_1 commutes with s_i and t_i for all positive integers i.

Proposition 14.20. If D is a countable, existentially complete division algebra over k and D_1 is a division subalgebra of D, then there is a division subalgebra D_1'' of D such that D_1 and D_1'' are isomorphic over k and $C_D(C_D(D_1'')) = D_1''$.

Proof. If D_1 is finitely generated, then $C_D(C_D(D_1)) = D_1$. Suppose that D_1 is not finitely generated. Construct D_ω as above. Let $\{a_i : i = 1, 2, \ldots\}$ be an enumeration of D_1, and let B_n be the division subalgebra

$$(\ldots((((\langle a_1, \ldots, a_n\rangle(s_1))(t_1; f_1))(s_2))(t_2; f_2)) \ldots (s_n))(t_n; f_n).$$

The division algebra B_n can be described up to isomorphism in an existentially complete division algebra by an existential sentence with parameters $a_1, \ldots, a_n$, so D includes a copy B'_n of B_n for each positive integer n. Since D is finitely homogeneous, one may assume that $B'_1 \subseteq B'_2 \subseteq \ldots$ and that the restriction to B_n of the natural isomorphism of B_{n+1} onto B'_{n+1} is the natural isomorphism of B_n onto B'_n. Then $D'_\omega = \bigcup_{0<n<\omega} B'_n$ and $D_\omega = \bigcup_{0<n<\omega} B_n$ are isomorphic.

Let D'_1 be the image of D_1 under the isomorphism of D_ω onto D'_ω, and let C' be the center of D'_1. The division subalgebra E' of D'_ω generated by C' and $\{s'_i, t'_i : i = 1, 2, \ldots\}$ is not included in any finitely generated division subalgebra of D'_ω. Therefore, D includes a copy $E'' \subseteq D''_\omega$ of $E' \subseteq D'_\omega$ such that $C_D(E'') = C_{D''_\omega}(E'')$ (apply Proposition 14.14). Let D''_1 be the image of D'_1 in D''_ω. The image C'' of C' is the center of D''_1. Since each element of D''_1 commutes with each element of $C'' \cup \{s''_i, t''_i : i = 1, 2, \ldots\}$, E'' is included in $C_D(D''_1)$. Consequently, $D''_1 \subseteq C_D(C_D(D''_1)) \subseteq C_D(E'') = C_{D''_\omega}(E)$ $\subseteq C_{D''_\omega}(\{s''_i, t''_i : i = 1, 2, \ldots\}) = D''_1$. Thus, $C_D(C_D(D''_1)) = D''_1$.

Corollary 14.21. If D is a countable, existentially complete division algebra over k and D_1 is a division subalgebra of D, then there are division subalgebras D'_1 and D'_2 of D such that D_1 and D'_1 are isomorphic over k and $C_D(D'_2) = D'_1$.

Proposition 14.20 cannot be improved for nonfinitely generated division subalgebras. Indeed, suppose D_1 is not finitely generated. Consider the division algebra $D_1(s)$ where s is a commuting indeterminate. The division algebra D must include a copy of $D_1(s)$. Since D_1 is not included in any finitely generated division subalgebra of $D_1(s)$, D includes a copy $D'''_1 \subseteq D'''_1(s''')$ of

$D_1 \subseteq D_1(s)$ for which $C_D(D_1''') = C_{D_1'''(s''')}(D_1''') = C'''(s''')$ where C''' is the center of D_1'''. Then $C_D(C_D(D_1''')) = C_D(C'''(s''')) \supseteq D_1'''(s''') \supsetneqq$ so D_1''' is not its own double centralizer.

Consider now the special case of centralizers of subfields of a countable, existentially complete division algebra over k. The possibilities for the centralizer of a subfield which is not finitely generated over k are quite unrestricted, as Corollary 14.15 shows. On the other hand, the centralizer of a finitely generated subfield is uniquely determined by the division algebra, because existentially complete division algebras are finitely homogeneous. While one cannot explicitly describe the centralizer, one can prove the following fact about the centralizer.

Proposition 14.22. If D is an existentially complete division algebra over k and K is a subfield which is finitely generated over k, then the centralizer of K in D is an existentially complete division algebra over K.

Proof. Let $D' = C_D(K)$. The field K is included in the center of D'. Suppose $\exists v_0 \ldots \exists v_n\ \psi(v_0, \ldots, v_n)$, where ψ is quantifier-free, is an existential sentence which is defined in D' and true in some extension D'' which is a division algebra over K. Let $e_0, \ldots, e_m$ be generators of K over k. The division algebra D'' satisfies the sentence

$$\exists v_0 \ldots \exists v_n\ (\psi(v_0, \ldots, v_n) \wedge \bigwedge_{i=0}^{m} \bigwedge_{j=0}^{n} (v_i e_j = e_j v_i)).$$

Let D''' be the amalgamation of D'' with D over k with amalgam D'. The sentence above is true in D'''. This sentence must hold in D, since D is existentially complete in D'''. Therefore, the sentence $\exists v_0 \ldots \exists v_n\ \psi(v_0, \ldots, v_n)$ is true in $C_D(K)$. Hence, $C_D(K)$ is existentially complete as a division algebra over K.

§ 4 Embeddings and Extensions

The principal results in this section are that each countable, existentially complete division algebra has a proper endomorphism, an automorphism which is not inner, and an existentially complete extension of cardinality $\aleph_1$ which does not realize any new existential types.

The following formulation for division algebras of a well-known property of homogeneous structures will be required for subsequent results.

Lemma 14.23. (i) Two finitely homogeneous division algebras D and D' over k have the same finitely generated division subalgebras if and only if $D \equiv_{\infty,\omega} D'$.

(ii) Assume that D and D' are finitely homogeneous division algebras over k and that D' extends D. Then D and D' have the same finitely generated division subalgebras if and only if $D \prec_{\infty,\omega} D'$.

(iii) A countably generated division algebra E over k can be embedded in a finitely homogeneous division algebra D over k if and only if each finitely generated division subalgebra of E is isomorphic to a division subalgebra of D.

(iv) Two finitely homogeneous, countably generated, division algebras over k are isomorphic if and only if each finitely generated division subalgebra of one is isomorphic to a division subalgebra of the other.

Proof. (i) It suffices to show that if $a_0, \ldots, a_n, b$ are elements of D and $a'_0, \ldots, a'_n$ are elements of D' for which $\langle a_0, \ldots, a_n\rangle$ and $\langle a'_0, \ldots, a'_n\rangle$ are isomorphic in the natural way, then there is an element b' of D' such that $\langle a_0, \ldots, a_n, b\rangle$ and $\langle a'_0, \ldots, a'_n, b'\rangle$ are isomorphic in the natural way (Karp's criterion (51) and the symmetry of the assumptions on D and D'). By hypothesis, there are elements $a_0'', \ldots, a_n''$, b'' of D' such that $\langle a_0, \ldots, a_n, b\rangle$ and $\langle a_0'', \ldots, a_n'', b''\rangle$ are isomorphic. Since D' is finitely homogeneous, there is an automorphism f of D' such that $f(a_i'') = a_i'$ for $i = 0, \ldots, n$. Let $b' = f(b'')$.

(ii) Since the division subalgebra generated by elements $a_0, \ldots, a_n$ of D is the same in both D and D', it suffices to show that for each b' in D' there is an element b in D such that $\langle a_0, \ldots, a_n, b\rangle$ and $\langle a_0, \ldots, a_n, b'\rangle$ are isomorphic. The existence of such an element b follows from the hypotheses on subalgebras and homogeneity as in the proof of part (i).

(iii) Let $\{e_i : i < \omega\}$ be a set of generators for E over k. By assumption, there is an element d_0 of D such that $\langle e_0\rangle$ and $\langle d_0\rangle$ are isomorphic. Suppose elements $d_0, \ldots, d_n$ of D have been chosen such that $\langle e_0, \ldots, e_n\rangle$ and $\langle d_0, \ldots, d_n\rangle$ are isomorphic. By hypothesis, there are elements $d_0', \ldots, d_n', d_{n+1}'$ of D such that $\langle e_0, \ldots, e_n, e_{n+1}\rangle$ and $\langle d_0', \ldots, d_n', d_{n+1}'\rangle$ are isomorphic. The finite homogeneity of D implies the existence of an element d_{n+1} of D such that $\langle d_0, \ldots, d_n, d_{n+1}\rangle$ and $\langle d_0', \ldots, d_n', d_{n+1}'\rangle$ are isomorphic. By induction, one obtains a collect $\{d_i : i < \omega\}$ of elements of D which generates a division subalgebra isomorphic to E.

(iv) The proof is similar to the proof of part (iii) with the addition of a back and forth construction.

Corollary 14.24. Parts (i) - (iv) of Lemma 14.23 hold with "finitely homogeneous" replaced by "existentially complete".

The next proposition is the basis for the other results in this section.

Proposition 14.25. If D is a countable, existentially complete division algebra over a countable field k and t is a transcendental element of D, then there is an embedding over k of D into $C_D(t)$.

<u>Proof</u>. Let $\{d_i : i < \omega\}$ be an enumeration of the elements of D. Let t' be an indeterminate which commutes with all elements of D. Embed $D(t')$ in an existentially complete division algebra E over k. Suppose elements $d'_0, \ldots, d'_n$ in $C_D(t)$ have been chosen such that $\langle d_0, \ldots, d_n \rangle (t')$ and $\langle d'_0, \ldots, d'_n \rangle (t)$ are isomorphic over k in the natural way. Since $k(t)$ and $k(t')$ are isomorphic over k, there is an inner automorphism of E which maps t' to t. Let $d''_0, \ldots, d''_{n+1}$ be the images of $d_0, \ldots, d_{n+1}$ under this mapping. The division subalgebra $\langle d''_0, \ldots, d''_{n+1} \rangle (t)$ is determined up to isomorphism in E by the following sentence $\phi(v_0, \ldots, v_{n+1})$:

$$\exists x \exists y \exists z (\sim(x = 0) \wedge \sim(y = 0) \wedge \sim(z = 0) \wedge \bigwedge_{i=0}^{n+1} (v_i t = t v_i$$

$$\wedge\, d_i x = x v_i \;\wedge\; v_i y = y v_i \;\wedge\; v_i z = z v_i)$$

$$\wedge\, ty = yt^2 \;\wedge\; t^2 z = zt^2 \;\wedge\; \sim(tz = zt)).$$

The sentence $\phi(d''_0, \ldots, d''_{n+1})$ is true in E. Since D is existentially complete in E, there are elements $d'''_0, \ldots, d'''_{n+1}$ in D for which D satisfies $\phi(d'''_0, \ldots, d'''_{n+1})$. The division subalgebras $\langle d'_0, \ldots, d'_n \rangle (t)$ and $\langle d'''_0, \ldots, d'''_n \rangle (t)$ are isomorphic, so there is an inner automorphism of D which maps t to t and d'''_i to d'_i for $i = 0, \ldots, n$. Let d'_{n+1} be the image of d'''_{n+1} under this mapping. Then the division subalgebras $\langle d_0, \ldots, d_{n+1} \rangle (t')$ and $\langle d'_0, \ldots, d'_{n+1} \rangle (t)$ are isomorphic.

This procedure yields a subset $\{d'_n : n < \omega\}$ of $C_D(t)$ which generates a division subalgebra isomorphic with D.

The preceding result is false if the hypothesis that t is transcendental is omitted. To verify this, suppose a is an element of D which is algebraic over k but is not in k, and let n be the number of distinct zeroes in an algebraic closure of k of the irreducible polynomial of a over k. The division algebra D satisfies an existential sentence which asserts that the irreducible polynomial of a over k has n distinct zeroes which commute with one another but none of which is in the center. The division subalgebr $C_D(a)$ cannot satisfy this sentence, so D cannot be embedded in $C_D(a$

Theorem 14.26. If D is a countable, existentially complete divi algebra over a countable field k, then D has a proper endomorphism such that $f(D) \prec_{\infty,\omega} D$.

Proof. This result is a consequence of Theorem 14.25 and Corollary 14.24.

Proposition 14.27. If D is a countable, existentially complete division algebra over a countable field k, then D has an extension of cardinality $\aleph_1$ such that $D \prec_{\infty,\omega} E$.

Proof. Construct an ascending chain $\{D_\alpha : \alpha < \omega_1\}$ of division algebras over k such that D_α is isomorphic to D for each α as follows. Let $D_0 = D$. Assume D_β has been constructed for each $\beta <$ If α is a successor ordinal, say $\alpha = \beta + 1$, then D_β has a proper endomorphic image $\bar{D}_\beta$. Let D_α be an extension of D_β such that D_β is an isomorphic copy of $\bar{D}_\beta \subseteq D_\beta$. Then D_α is isomorphic to D_β ar so is isomorphic to D. If α is a limit ordinal, then let $D_\alpha = \bigcup_{\beta<\alpha}$ The algebra D_α is countable, existentially complete, and each finite generated division subalgebra of D_α is included in some D_β with $\beta < \alpha$. Therefore, D_α is isomorphic to D (Corollary 14.24).

Let $E = \bigcup_{\alpha<\omega_1} D_\alpha$. The division algebra E is an existentially complete division algebra over k of cardinality $\aleph_1$. Also, E includes D, and a copy of each finitely generated division subalgebra of E occurs in D. Hence, according to Corollary 14.24, $D \prec_{\infty,\omega} E$.

Corollary 14.28. If D is a countable, existentially complete division algebra over a countable field k, then D has $2^{\aleph_0}$ distin automorphisms. In particular, D has an automorphism which is not inn

Proof. This can be inferred directly from the preceding result and a result of Keuker (54). Alternatively, one can construct $2^{\aleph_0}$ distinct automorphisms directly from Lemma 14.3 and Corollary 14.7.

Corollary 14.29. If k is a countable field, then there is a finitely generic division algebra over k of cardinality $\aleph_1$.

Proof. The finitely generic structures are axiomatized by a sentence of $\mathcal{L}(k)_{\omega_1,\omega}$. Therefore, if D is a countable, finitely generic division algebra over k, then the division algebra E constructed in Proposition 14.27 is finitely generic also.

Whether finitely generic division algebras over k of arbitrarily large cardinality exist is an open question. A. Macintyre (59) has shown that Martin's axiom implies the existence of finitely generic division algebras over $\mathbb{Q}$ or F_p in all cardinalities not exceeding $2^{\aleph_0}$.

Theorem 14.12 implies that there are no minimal existentially complete division algebras over k. A. Macintyre (60) has shown that there are no prime existentially complete division rings of characteristic 0 or p, that is, existentially complete division algebras which can be embedded in every existentially complete division ring of the same characteristic.

5 The Number of Finitely Generated Division Algebras

Uncountably many nonisomorphic finitely generated division algebras will be constructed in this section. In turn, the existence of these division algebras will be used in the proof that each existentially universal division algebra is uncountable. The method of construction of these finitely generated division algebras is relevant to the constructions in Chapter 16 also.

Theorem 14.30. There are $2^{\aleph_0}$ pairwise nonisomorphic division algebras generated over k by four elements.

Proof. B. H. Neumann (68) constructed $2^{\aleph_0}$ pairwise nonisomorphic permutation groups generated by two elements. The strategy here is to realize these permutations as inner automorphisms of division algebras.

Let $K = k(t_0, t_1, \ldots)$ be a purely transcendental extension field of k generated by a set $\{t_i : i < \omega\}$ of commuting, algebraically independent elements. The field K has a proper endomorphism defined by $\gamma(t_i) = t_{i+1}$ for all $i < \omega$.

Let $s_{i,j} = t_{i(i+2)+j}$ for $0 \leq j \leq 2 + 2i$. This partitions the t into a sequence of sets $S_i = \{s_{i,j} : 0 \leq j \leq 2 + 2i\}$ where S_i has $2i + 3$ elements. Suppose $\mathcal{O}$ is an infinite set of odd integers greater than 1. Define mappings $\alpha_{\mathcal{O}} : K \to K$ and $\beta_{\mathcal{O}} : K \to K$ by

$$\alpha_{\mathcal{O}}(s_{i,j}) = \begin{cases} s_{i,j} & \text{if } 2_i + 3 \in \mathcal{O} \\ s_{i,j+1} & \text{if } 2i + 3 \in \mathcal{O} \text{ and } j < 2i + 2 \\ s_{i,0} & \text{if } 2_i + 3 \in \mathcal{O} \text{ and } j = 2i + 2. \end{cases}$$

$$\beta_{\mathcal{O}}(s_{i,j}) = \begin{cases} s_{i,j} & \text{if } 2i + 3 \notin \mathcal{O} \\ s_{i,j} & \text{if } 2i + 3 \in \mathcal{O} \text{ and } j > 2 \\ s_{i,1} & \text{if } 2i + 3 \in \mathcal{O} \text{ and } j = 0 \\ s_{i,2} & \text{if } 2i + 3 \in \mathcal{O} \text{ and } j = 1 \\ s_{i,0} & \text{if } 2i + 3 \in \mathcal{O} \text{ and } j = 2. \end{cases}$$

Neumann (68) proved that the group $G_{\mathcal{O}}$ of permutations of the set $\{t_i : i < \omega\}$ generated by $\alpha_{\mathcal{O}}$ and $\beta_{\mathcal{O}}$ is determined uniquely up to isomorphism by $\mathcal{O}$.

There is a division algebra D over k which includes K and in which the mappings γ, $\alpha_{\mathcal{O}}$, and $\beta_{\mathcal{O}}$ are induced by inner automorphisms corresponding to elements c, $a_{\mathcal{O}}$, and $b_{\mathcal{O}}$, respectively Let $D_{\mathcal{O}}$ be the division subalgebra generated by t_0, $a_{\mathcal{O}}$, $b_{\mathcal{O}}$, and c. The algebra $D_{\mathcal{O}}$ includes K, since $c^{-n}t_0c^n = t_n$ for $n \geq 1$. Also, $a_{\mathcal{O}}$ and $b_{\mathcal{O}}$ induce the permutations $\alpha_{\mathcal{O}}$ and $\beta_{\mathcal{O}}$.

If D' is a division algebra over k generated by four elements, let $\mathcal{H}_{D'}$ be the collection of four-tuples (a, b, c, d) of elements of D' for which (i) D' is generated over k by a, b, c, d, (ii) $\langle c^{-n}dc^n : n < \omega\rangle$ is isomorphic over k to K in the natural way, and (iii) the inner automorphisms determined by a and b induce permutations on the sets $\overline{S}_i = \{c^{-n}dn : t_n \in S_i\}$. If (a, b, c, d) is an element in $\mathcal{H}_{D'}$, then the inner automorphisms determined by a and b generate a group of permutation of $\{c^{-n}dc^n : n < \omega\}$. Let $\mathcal{B}_{D'}$ be the collection of these groups. The set $\mathcal{B}_{D'}$ is countable, because D' is generated by four elements. Since there are $2^{\aleph_0}$ groups of the form G_θ and G_θ is a member of $\mathcal{B}_{D_\theta}$, there must be $2^{\aleph_0}$ nonisomorphic division algebras generated over k by four elements.

Corollary 14.31. No existentially universal division algebra over k is countable. Moreover, if k is countable, then the class of existentially universal division algebras over k is a proper subclass of the infinitely generic division algebras over k.

Proof. Each existentially universal division algebra over k must include a copy of every finitely generated division algebra over k and so must have cardinality $2^{\aleph_0}$. However, if k is countable, then there do exist countable, infinitely generic division algebras over k.

Corollary 14.32. There are $2^{\aleph_0}$ pairwise nonisomorphic division algebras generated over k by two elements.

Proof. This is a consequence of Theorem 14.30 and a result of P. M. Cohn that each division algebra countably generated over k can be embedded in a division algebra generated over k by two elements.

The preceding corollary for $k = \mathbb{Q}$ or $k = F_p$, the finite field with p elements, was obtained independently by A. Macintyre using Cohn's result and the existence of $2^{\aleph_0}$ pairwise nonisomorphic, finitely generated, ordered groups.

Analogues of most results in this chapter are true for groups. A correct analogue is produced usually through replacing "division algebra" by "group", "division subalgebra" by "subgroup", "transcendental" by "infinite order", and so forth. Proofs of the analogues are either the same or easier. Specific comments on each section follow.

Theorem 14.1 of section one corresponds to a well-known theorem of G. Higman, B. H. Neumann, and H. Neumann (45) on isomorphisms of subgroups. Indeed, their result motivated Theorem 14.1. Lemma 14.2 and Lemma 14.3 hold for finitely generated subgroups of existentially complete groups (see A. Macintyre (56)). Analogues of results 14.4 through 14.7 are true for groups also. Proposition 14.8 has no appropriate analogue for groups.

The formula $\text{Transc}_n(v_1, \ldots, v_n)$ in section two is transformed into a formula of group theory by the omission of conjuncts which assert that elements are nonzero. The analogue for groups of a collection of commuting, algebraically independent elements is a collection of commuting elements which freely generate a free, abelian group. The resultant analogue of Lemma 14.9 is true for existentially complete groups. The correct analogues for groups of a countably generated field, an infinite, transcendence basis, and an algebraically closed field are a countably generated abelian group, an infinite set of commuting, free generators, and an algebraically closed abelian group, respectively.

Correct analogues for groups of all propositions, theorems, and corollaries in section three are produced by replacing "maximal subfields" by "maximal abelian subgroups".

The analogue of Corollary 14.24 is due to A. Macintyre (56). A stronger result than Theorem 14.25 is true for groups. Namely, if G is a countable, existentially complete group and g is an element of G, then G can be embedded in $C_G(g)$. Analogues of all other results in section four appear in A. Macintyre's article on algebraically closed groups (56).

Theorem 14.30 in section five corresponds to B. H. Neumann's theorem on two-generator groups. The analogue for groups of Corollary 14.31 is true also (A. Macintyre (56)).

CHAPTER 15

NULLSTELLENSATZ

The Nullstellensatz is an important theorem of commutative algebra and one of the foundations of algebraic geometry. This theorem asserts that whenever K is an algebraically closed field, I is an ideal in the polynomial ring $K[x_1,\ldots, x_n]$, and $g(x_1,\ldots, x_n)$ is a polynomial in $K[x_1,\ldots, x_n]$, then $g(x_1,\ldots, x_n)$ shares all common zeroes (in K) of I if and only if $g(x_1,\ldots, x_n)$ is in the radical of I. An analogue of this theorem is true also for existentially complete division algebras.

All rings mentioned in this chapter will be algebras over a field k. Ideals will be closed under the operation of k on the ring.

First, one must define noncommutative polynomials and a suitable radical. Let D be a division algebra over k and let $x_1, \ldots, x_n$ be indeterminates which commute only with elements of k. A polynomial over D in $x_1, \ldots, x_n$ is a sum of monomials of the form $a_1x_{i_1}a_2x_{i_2}a_3 \cdots a_nx_{i_n}a_{n+1}$, where the a_i are elements of D. Equivalently, a polynomial over D as an algebra over k is an element of the coproduct of D and the free algebra $k\langle x_1,\ldots, x_n\rangle$ in the category of algebras over k. This definition coincides with that of P. M. Cohn (25, 26) and, when k is the center of D, with that of S. Amitsur (2). Individual polynomials will be denoted by $g(x_1,\ldots, x_n)$ or $h(x_1,\ldots, x_n)$ with or without subscripts. The ring of polynomials over the division algebra D as an algebra over k will be denoted by $D_k[x_1,\ldots, x_n]$. This ring of noncommutative polynomials has the following properties: (i) if D and D' are division algebras over k and D' extends D, then $D'_k[x_1,\ldots, x_n]$ extends $D_k[x_1,\ldots, x_n]$; (ii) if R, an algebra over k, is finitely generated over a division subalgebra D, then

R is a homomorphic image of $D_k[x_1, \ldots, x_n]$ for some n; and (iii) $D_k[x_1, \ldots, x_n]$ can be embedded in a division algebra over k.

This definition of noncommutative polynomials has two admitted drawbacks. First, a nonzero polynomial may define the zero function. For example, the polynomial $ixi + x$ in the ring $\mathbb{C}_{\mathbb{R}}[x]$ is nonzero since x does not commute with i, but $ixi + x$ defines the zero function $f(x) = 0$ on $\mathbb{C}$. However, if D is an existentiall complete division algebra over k, then D is infinite dimensional over its center k, so no nonzero polynomial in $D_k[x_1, \ldots, x_n]$ defines the zero function on D (Amitsur (2)). Secondly, a nonconstant polynomial may have no zeroes in any division algebra over k. The following example is due to P. M. Cohn (26). Let k be a field whose algebraic closure $\bar{k}$ is not a purely inseparable extension of k. Let a be a separable element of $\bar{k} - k$. The polynomial $ax - xa - 1$ in $k(a)_k[x]$ has no solution in any division algebra over k. To verify this, suppose that b is a solution in some division algebra over k, and let p(x) be the irreducible polynomial of a over k. Then $p'(a) = p(a)b - bp(a) = 0 \cdot b - b \cdot 0 = 0$ contradicting that a is separable over k.

A division algebra D over k is existentially complete if and only if each finite set of polynomials over D which has a common zero in an extension of D has a common zero in D itself. The necessity of the latter condition is clear, since the existence of a common zero for a finite set of polynomials can be formulated as an existential sentence with paramaters from D. The sufficiency of the latter condition depends on the replacement of negated atomic formulas by positive existential formulas. Such a replacement is possible because the formula $\sim(\xi_i = \xi_j)$, where ξ_i and ξ_j are terms, is logically equivalent in the theory of division rings to the formula $\exists v(v \cdot \xi_i = v \cdot \xi_j + 1)$. Structures which satisfy conditions like the latter one are usually called <u>algebraically closed</u> (see P. Eklof and

G. Sabbagh (34), W. R. Scott (101)). Thus, a division algebra over k is existentially complete if and only if it is algebraically closed.

Next, one must define a suitable radical. The radical of an ideal in a commutative ring is the intersection of the prime ideals which contain the given ideal. The property of prime ideals on which the Nullstellensatz depends is that an ideal P of a commutative ring R is prime if and only if R/P can be embedded in a field. Accordingly, an ideal J in an algebra R over k will be called a d-prime ideal if R/J can be embedded into a division algebra over k by an algebra homomorphism. An ideal in a commutative ring is d-prime if and only if it is prime. However, an ideal in a noncommutative ring may be prime but not d-prime. For example, the zero ideal of a noncommutative ring which has no zero-divisors but cannot be embedded in a division ring (see Malcev (64)) is prime but not d-prime. The d-radical of an ideal I in an algebra R over k is the intersection of all d-prime ideals of R which include I. The d-radical of R is the d-radical of the zero ideal. Since R considered as an ideal is d-prime, every ideal has a d-radical.

Not surprisingly, the elements of the d-radical of an ideal are no longer just those elements for which some power lies in the ideal. However, the elements of the d-radical can be characterized without reference to d-prime ideals. Since this characterization is not necessary for the proof of the Nullstellensatz, it will be postponed until the end of this chapter.

Theorem 15.1 (Nullstellensatz). Assume that D is an existentially complete division algebra over k. Suppose that $g(x_1,\dots, x_n)$ is a polynomial in $D_k[x_1,\dots, x_n]$ and that I is an ideal in $D_k[x_1,\dots, x_n]$.

(i) If $g(x_1,\dots, x_n)$ shares all zeroes of I in every division algebra over k extending D, then $g(x_1,\dots, x_n)$ is in the d-radical of I. If, in addition, I is finitely generated, then only zeroes of

I in D need by considered.

(ii) Conversely, if $g(x_1,\ldots, x_n)$ is in the d-radical of I, then every zero of I is also a zero of $g(x_1,\ldots, x_n)$.

Proof. (i) Suppose $g(x_1,\ldots, x_n)$ is not in the d-radical of I. Then there is a d-prime ideal J which includes I but not $g(x_1,\ldots, x_n)$. Denote the equivalence class of a polynomial $h(x_1,\ldots, x_n)$ in $D_k[x_1,\ldots, x_n]/J$ by $\overline{h(x_1,\ldots, x_n)}$. Since $g(x_1,\ldots, x_n)$ is not in J, $0 \neq \overline{g(x_1,\ldots, x_n)} = g(\bar{x}_1, \bar{x}_2, \ldots, \bar{x}_n)$. Embed $D_k[x_1,\ldots, x_n]/J$ in a division algebra D' over k by a monomorphism f. Since $(D + J)/J$ is isomorphic to D, D may be identified with $f((D + J)/J)$ in D'. Then $0 \neq f(\overline{g(x_1,\ldots, x_n)}) = f(g(\bar{x}_1,\ldots, \bar{x}_n)) = g(f(\bar{x}_1),\ldots, f(\bar{x}_n))$. If $h(x_1,\ldots, x_n)$ is a polynomial in I, then it is in J, so $h(f(\bar{x}_1), \ldots, f(\bar{x}_n)) = f(h(\bar{x}_1,\ldots, \bar{x}_n)) = f(\overline{h(x_1,\ldots, x_n)}) = f(0) = 0$. Thus, the elements $f(\bar{x}_1), \ldots, f(\bar{x}_n)$ in D' are a zero of I but not a zero of $g(x_1,\ldots, x_n)$, contrary to hypothesis.

Suppose, in addition, that I is finitely generated, say by $h_1(x_1,\ldots, x_n), \ldots, h_m(x_1,\ldots, x_n)$. Then D' satisfies the formula

$$\exists v_1 \ldots \exists v_n(h_1(v_1,\ldots, v_n) = 0 \wedge \ldots \wedge h_m(v_1,\ldots, v_n) = 0 \wedge \sim(g(v_1,\ldots, v_n) = 0)).$$

D must satisfy this sentence also, because D is existentially complete. Consequently, there is a zero of I in D which is not a zero of $g(x_1,\ldots, x_n)$, contrary to hypothesis again.

(ii) Conversely, suppose $g(x_1,\ldots, x_n)$ is in the d-radical of I. Suppose that D' is a division algebra over k extending D and that elements $d_1,\ldots, d_n$ of D' are a zero of I. There is a homomorphism $f : D_k[x_1,\ldots, x_n] \to D'$ defined by $x_i \to d_i$ for $i = 1, \ldots, n$. The kernel of f is a d-prime ideal including I. The polynomial $g(x_1,\ldots, x_n)$ is in the kernel of f, since $g(x_1,\ldots, x_n)$ is in the d- radical of I, so $0 = f(g(x_1,\ldots, x_n)) = g(f(x_1), \ldots, f(x_n)) = g(d_1,\ldots, d_n)$.

Part (i) of the Nullstellensatz cannot be improved in general.

Proposition 15.2. If k is a countable field and D is a countable, existentially complete division algebra over k, then there is an infinitely generated ideal in $D_k[x]$ and a polynomial $g(x)$ in $D_k[x]$ such that $g(x)$ shares all zeroes of I in D but is not in the d-radical of I.

Proof. The algebra D has an automorphism f which is not inner, according to Corollary 14.28. Let I be the ideal of $D_k[x]$ generated by $\{dx - xf(d) : d \in D\}$. The only zero of I in D is 0, since f is not inner. However, I does have a nontrivial zero in some extension, because f is an inner automorphism in some extension (Theorem 14.1). Let $g(x) = x$. The only zero of $g(x)$ is 0. Thus $g(x)$ shares all zeroes of I in D but does not share all zeroes of I in all extensions of D. Consequently, $g(x)$ is not in the d-radical of I.

Corollary 15.3. With the hypotheses above, the algebra $D_k[x_1, \ldots, x_n]$ is not Noetherian.

Proof. The ideal I in the preceding proof cannot be finitely generated.

The Nullstellensatz for division algebras may be interpreted geometrically as in the commutative case. The variety $V(S)$ of a subset S of $D_k[x_1, \ldots, x_n]$ is the set of points in the n-fold cartesian product D^n which are zeroes of S, that is,

$$V(S) = \{(d_1, \ldots, d_n) \in D^n : g(d_1, \ldots, d_n) = 0 \text{ for all } g(x_1, \ldots, x_n) \in S\}.$$

A subset A of D^n is called a variety if $A = V(S)$ for some subset S of $D_k[x_1, \ldots, x_n]$. The ideal $I(A)$ of a subset A of D^n is the set $\{g(x_1, \ldots, x_n) \in D_k[x_1, \ldots, x_n] : g(d_1, \ldots, d_n) = 0$ for all $(d_1, \ldots, d_n) \in A\}$. An ideal in $D_k[x_1, \ldots, x_n]$ is said to be d-radical if it is its own d-radical.

Theorem 15.4. Let D be an existentially complete division algebra over k.

(i) Two d-radical ideals in $D_k[x_1,\dots, x_n]$ are distinct if and only if they have different varieties in some division algebra over k extending D.

(ii) If A is a variety in $(D')^n$ for a division algebra D' ov k extending D and I'(A) is the ideal of A in $D'_k[x_1,\dots, x_n]$, then $I'(A) \cap D_k[x_1,\dots, x_n]$ is a d-radical ideal in $D_k[x_1,\dots, x$

Proof. Part (i) follows from Theorem 15.1. To prove part (ii), let $I = I'(A) \cap D_k[x_1,\dots, x_n]$, and suppose that $g(x_1,\dots, x_n)$, a polynomial in $D_k[x_1,\dots, x_n]$, is not in I. Then there are elements $d_1', \dots, d_n'$ in D' such that $(d_1',\dots, d_n')$ is a point of A and $g(d_1',\dots, d_n') \neq 0$. Since $(d_1',\dots, d_n')$ is a zero of I, $g(x_1,\dots, x_n)$ is not in the d-radical of I. Therefore, I coincides with its d-radical.

These results may contribute to the development of a noncommutativ algebraic geometry (see Cohn (25, 26, 28), Procesi (73, 74)).

As previously mentioned, the elements of the d-radical of an ideal can be characterized without reference to the intersection of the d-pri ideals including that ideal. In fact, there are two characterizations, a metamathematical one and an algebraic one.

The metamathematical characterization is a generalization of a theorem of A. Robinson (78). Let R be an algebra over k and let I be an ideal of R. Add names for all elements of I to the languag of T_k to form a new language $\mathcal{L}(I)$. Let I^+ be the positive diagram of I in the language $\mathcal{L}(I)$. Let S_I be the set of formulas $\phi(v_0,\dots, v_n)$ of the language $\mathcal{L}(I)$ such that $\phi(v_0,\dots, v_n)$ is a conjunction of atomic formulas and

$$T_k \cup I^+ \cup \{a = 0 : a \in I\} \vdash \forall v_0 \dots \forall v_n(\phi(v_0,\dots, v_n) \to v_0 = 0).$$

Proposition 15.5. The set $J_I = \{r \in R :$ there are elements $r_1, \ldots, r_n$ in R and a formula $\phi(v_0, \ldots, v_n)$ in S_I such that R satisfies $\phi(r, r_1, \ldots, r_n)\}$ is the d-radical of I. In particular, J_0 is the d-radical of R.

Proof. First, suppose r is an element of J_I. There are elements $r_1, \ldots, r_n$ of R and a formula $\phi(v_0, \ldots, v_n)$ in S_I for which R satisfies $\phi(r, r_1, \ldots, r_n)$. Let J be a d-prime ideal which includes I. Since J is d-prime, there is a division algebra D over k and a monomorphism $f : R/J \to D$. Denote the image of an element r of R under the mapping $R \to R/J \overset{f}{\to} D$ by $\bar{r}$. Assign each constant a of $\mathcal{L}(I)$ to name the image $\bar{a}$ in D. Then D must satisfy $\phi(\bar{r}, \bar{r}_1, \ldots, \bar{r}_n)$, since $\phi(v_0, \ldots, v_n)$ is positive and R satisfies $\phi(r, r_1, \ldots, r_n)$. But D is also a model of $T_k \cup I^+ \cup \{a = 0 : a \in I\}$, so $\bar{r} = 0$. Thus r is in J.

Conversely, suppose r is an element of $R - J_I$. Let R^+ be the positive diagram of R. One of two cases holds.

(i) $T_k \cup R^+ \cup \{a = 0 : a \in I\} \cup \{\sim(r = 0)\}$ is consistent. Then there is a division algebra D over k which is a model of this set of sentences. Let R' be the subalgebra of D generated by elements named in R^+. There is a surjective homomorphism $f : R \to R'$, since R' is a model of R^+. Since R' is a model of $\{a = 0 : a \in I\}$, I is included in the kernel of f, which is a d-prime ideal. But r is not in the kernel of f. Thus, r is not in the d-radical of I.

(ii) $T_k \cup R^+ \cup \{a = 0 : a \in I\} \cup \{\sim(r = 0)\}$ is inconsistent. Then there is a conjunction ϕ of formulas in R^+ such that $T_k \cup I^+ \cup \{a = 0 : a \in I\} \vdash \phi \to (r = 0)$. Replacing distinct constants which name elements of $R - I$ by distinct variables in ϕ yields a formula in S_I. This contradicts that r is not in J_I.

The algebraic characterization of the d-radical is due to P. M. Cohn (22). This characterization is related to properties of determinants in commutative algebra. The *diagonal sum* of two square matrices A and B, denoted by $A \dot{+} B$, is the matrix

$$\begin{pmatrix} A & 0 \\ 0 & B \end{pmatrix}$$

If A and B are square matrices of the same size which differ only in the first row, then their *determinantal sum*, denoted by $A \nabla B$, is the matrix

$$\begin{pmatrix} a_{1,1} + b_{1,1} & a_{1,2} + b_{1,2} & \cdots & a_{1,n} + b_{1,n} \\ a_{2,1} & a_{2,2} & \cdots & a_{2,n} \\ \vdots & \vdots & & \vdots \\ a_{n,1} & a_{n,2} & \cdots & a_{n,n} \end{pmatrix},$$

where A has entries $a_{i,j}$ and B has entries $b_{i,j}$. The determinantal sum of A and B is defined similarly whenever A and B differ in at most one row or column. The operation of forming determinantal sums is not associative.

An $n \times n$ matrix A is said to be *nonfull* if there exist an $n \times r$ matrix B and an $r \times n$ matrix C with $r < n$ such that $A = B$
Otherwise, A is said to be *full*.

Note that if A and B are matrices with entries from a commutati
ring, then $\det(A \dot{+} B) = \det(A) \cdot \det(B)$, $\det(A \nabla B) = \det(A) + \det(B)$, and if A is nonfull, then $\det(A) = 0$.

The *matrix ideal* generated by a set of square matrices with entries from a ring R is the smallest set W of square matrices with entries from R which satisfies

(i) W includes all nonfull matrices;

(ii) if A is in W and B is any square matrix, then $A \dot{+} B$ is in W;

(iii) if A and B are in W and $A \nabla B$ is defined, then

$A \nabla B$ is in W; and

(iv) if $A \dotplus I$ is in W for any identity matrix I, then A is in W.

Proposition 15.6. An element r is in the d-radical of an ideal I in R if and only if a finite diagonal sum of the matrix (r) with itself is in the matrix ideal generated by $\{(a) : a \in I\}$.

Proof. See P. M. Cohn (22, 27).

CHAPTER 16

CLASSES OF EXISTENTIALLY COMPLETE DIVISION ALGEBRAS

Several questions concerning classes of existentially complete structures arise from the theory developed in Part One (see the summary at the end of Part One). A theory has a model-companion if and only if at least one of the classes $\mathcal{E}_T$, $\mathcal{F}_T$, $\mathcal{G}_T$, or $\mathcal{A}_T$ is a generalized elementary class, in which case $\mathcal{E}_T = \mathcal{F}_T = \mathcal{G}_T$ is a generalized elementary class. However, the theory T_k of division algebras over a field k does not have a model-companion (Theorem 14.13). The absence of a model-companion does not imply that the classes $\mathcal{E}_T$, $\mathcal{G}_T$ and $\mathcal{F}_T$ are distinct. Indeed, if R is a noncoherent ring, then the theory of right R-modules has no model-companion but the classes $\mathcal{E}_T$, $\mathcal{G}_T$, and $\mathcal{F}_T$ coincide nevertheless. That the classes $\mathcal{E}_{T_k}$, $\mathcal{G}_{T_k}$, and $\mathcal{F}_{T_k}$ are distinct will be shown in this chapter. Other results in this chapter concern the number of theories of existentially complete division algebras over k, the existence of sentences of low quantifier complexity which are in one of T_k^F or T_k^f but not both, and the degree of unsolvability of the diagrams of some existentially complete division algebras.

This chapter consists of three sections. Structures for second order arithmetic which are arithmetically definable in existentially complete division algebras are introduced in the first section. In the second section, second order arithmetic is interpreted within the language of division algebras over k. Finally, these second order structures for arithmetic and this interpretation of second order arithmetic are used in the third section to answer the questions posed above.

§ 1 Structures for Second Order Arithmetic

The structures for second order arithmetic which are arithmetically definable within existentially complete division algebras over k will consist of the nonnegative powers of a transcendental element together with certain subsets of these powers.

Proposition 16.1. Let p be a prime distinct from the characteristic of the field k. The only rational functions $R(x)$ over k which satisfy $R(x^p) = (R(x))^p$ are those of the form ax^n, where a is a $(p-1)^{th}$ root of unity and n is an integer.

Proof. If $R(x) = ax^n$ where a is a $(p-1)^{th}$ root of unity, then $R(x^p) = a(x^p)^n = ax^{pn} = a^p x^{np} = (ax^n)^p$.

Conversely, assume $R(x)$ is a rational function over k which satisfies $R(x^p) = (R(x))^p$. One may assume without loss of generality that $R(x) = f(x)/g(x)$ where $f(x)$ and $g(x)$ are relatively prime polynomials, $f(x)$ is monic, and x divides neither $f(x)$ nor $g(x)$. Since $R(x^p) = (R(x))^p$, $1 = (R(x))^p \cdot \dfrac{1}{R(x^p)} = \dfrac{(f(x))^p}{(g(x))^p} \cdot \dfrac{g(x^p)}{f(x^p)}$. The polynomials $f(x^p)$ and $g(x^p)$ are relatively prime, since $f(x)$ and $g(x)$ are relatively prime. Consequently, $f(x^p)$ divides $(f(x))^p$. Also, $1 = R(x^p) \cdot \dfrac{1}{(R(x))^p} = \dfrac{f(x^p)}{g(x^p)} \cdot \dfrac{(g(x))^p}{(f(x))^p}$ and $(f(x))^p$ and $(g(x))^p$ are relatively prime, so $(f(x))^p$ divides $f(x^p)$. Both $f(x^p)$ and $(f(x))^p$ are monic, because $f(x)$ is monic, so $f(x^p) = (f(x))^p$. Then $1 = \dfrac{f(x^p)}{g(x^p)} \cdot \dfrac{(g(x))^p}{(f(x))^p} = \dfrac{(g(x))^p}{g(x^p)}$, so $g(x^p) = (g(x))^p$ also.

Thus, it suffices to show that if $h(x)$ is a polynomial which is not divisible by x and satisfies $h(x^p) = (h(x))^p$, then $h(x) = a \in k$ where $a^{p-1} = 1$. Since $h(x)$ is not divisible by x, $h(x) = a_0 + a_1 x + \dots + a_r x^r$ where $a_0 \neq 0$. The equality $h(x^p) = (h(x))^p$ implies $a_0 = a_0{}^p$, so $1 = a^{p-1}$. Suppose $h(x) \neq a_0$. Let i be the least positive subscript for which $a_i \neq 0$. The nonzero term of least positive degree in $h(x^p)$ is $a_i x^{pi}$. The term of least positive degree in $(h(x))^p$ is $pa_0 a_i x^i$, since p is not the characteristic of k and $a_0 \neq 0$ and $a_i \neq 0$. But pi is greater than i, so this contradicts that $h(x^p) = (h(x))^p$. Hence $h(x) = a_0$, which is a $(p - 1$ root of unity.

Corollary 16.2. The only rational functions $R(x)$ over a field k which satisfy $R(x^2) = (R(x))^2$ and $R(x^3) = (R(x))^3$ are the positive, zero, and negative powers of x.

Proof. Either 2 or 3 is not the characteristic of k, so $R(x) = ax^n$ where $a^{p-1} = 1$ either for $p = 2$ or for $p = 3$. Since $R(x^2) = (R(x))^2$, $ax^{2n} = a^2 x^{2n}$, so $a = 1$.

Let Power(x,y) be the formula

$$\sim(y=0) \wedge \forall z((xz = zx \to yz = zy) \wedge (xz = zx^2 \to yz = zy^2) \wedge (xz = zx^3 \to yz = zy^3)).$$

If t and s are elements of an existentially complete division algebra D over a field k and if t is transcendental over k, then D satisfies Power(t, s) if and only if s is a power of t. This is a consequence of Lemma 14.3, the preceding corollary, and the fact that $k(t)$ is isomorphic to the field of rational functions in one indeterminate.

Let PosPower(x,y) be the formula

$$\text{Power}(x,y) \wedge \forall z(zx = xz^2 \to \forall w_1 \forall w_2((z^2 w_1 = w_1 z^2 \wedge zy = yw_2) \to w_1 w_2 = w_2 w_1)) .$$

Lemma 16.3. The sentence PosPower(t, s) is satisfied in an existentially complete division algebra D over k by an element s and an element t which is transcendental over k if and only if s is a positive power of t.

Proof. The formula PosPower(t, s) holds for a transcendental element t and an arbitrary element s if and only if s is a power of t and $s^{-1}as$ is in the division subalgebra $\langle a^2 \rangle$ whenever $t^{-1}at = a^2$. If $s = t^n$ for some $n \geq 1$, then $t^{-1}at = a^2$ implies $s^{-1}as = a^{2n} \in \langle a^2 \rangle$, so PosPower(t, s) holds. Conversely, suppose D satisfies PosPower(t, s). Let a be an element of D which is transcendental over k. The division subalgebras $\langle a \rangle$ and $\langle a^2 \rangle$ are isomorphic, so there is an element c in D such that $c^{-1}ac = a^2$. Moreover, c may be chosen to be transcendental over k itself (see Theorem 14.1). Since D is finitely homogeneous, one may assume c is just t. There is an integer n such that $s = t^n$, because Power(t, s) holds. The equality $t^{-1}at = a^2$ implies $s^{-1}as = t^{-n}at^n = a^{2^n}$. If $n > 0$, then a^{2^n} is in $\langle a^2 \rangle$. If $n = 0$, then $a^{2^n} = a$ is not in division subalgebra $\langle a^2 \rangle$. Finally, if $n < 0$, say $n = -m$, then $a^{2^n} = a^{2^{-m}}$ is a 2^m root of a and so is not in the division subalgebra $\langle a^2 \rangle$. But by assumption $s^{-1}as$ is in $\langle a^2 \rangle$ because PosPower(t, s) holds. Hence, n is positive.

Let NNPower(x,y) be the formula PosPower(x,y) $\vee$ $y = 1$. An existentially complete division algebra over k satisfies NNPower(t, s) for an element s and an element t which is transcendental over k if and only if $s = t^n$ for some $n \geq 0$.

The following, arithmetically definable operations on the nonnegative powers of a transcendental element t yield a structure which is isomorphic to the nonnegative integers:

$$t^n \oplus t^m = t^q \leftrightarrow t^n \cdot t^m = t^{n+m} = t^q, \text{ and}$$

$$t^n \otimes t^m = t^q \leftrightarrow \forall z[(tz = zt^n \wedge \sim(z = 0)) \rightarrow t^m z = zt^q])$$

$$\leftrightarrow \exists z(tz = zt^n \wedge \sim(z = 0) \wedge t^m z = zt^q).$$

This completes the construction of a model of arithmetic which is arithmetically definable in each existentially complete division algebra.

This model is enlarged to a structure for second order arithmetic through the adjunction of subsets defined by pairs of inner automorphisms. Let Subset(x, y, z) be the formula $\sim(x = 0) \wedge \sim(y = 0) \wedge \sim(z = 0) \wedge \forall w(yxz = xzw \rightarrow \sim(yx = xw))$. The sentence Subset(a, b, c) is true for elements a, b, and c of a division algebra if and only if $(a^{-1}ba)c \neq c(a^{-1}ba)$, that is, $a^{-1}ba$ is not in the centralizer of c. Clearly, the following equivalence holds in any division algebra:

$$\text{Subset}(x, y, z) \leftrightarrow \sim(x = 0) \wedge \sim(y = 0) \wedge \sim(z = 0) \wedge \exists w\,(yxz = xzw \wedge \sim(yx = xw)).$$

If t is a transcendental element of an existentially complete division algebra D, then each pair (c, d) of elements of D determines a unique subset of the nonnegative powers of t, namely $\{t^n : \text{Subset}(t^n, c, d) \text{ holds in } D\}$. The nonnegative powers of t with the operations $\oplus$ and $\otimes$ together with the collection $\{\{t^n : D \text{ satisfies Subset}(t^n, c, d)\} : c, d \in D\}$ of subsets of powers of t constitutes a structure for second order arithmetic. The finite homogeneity of D implies that any two transcendental elements determine the same structure for second order arithmetic. Thus, each existentially complete division algebra D determines uniquely a structure $\mathcal{N}_D$ for second order arithmetic. These structures will be used to distinguish between members of $\mathcal{F}_T$, $\mathcal{G}_T$, $\mathcal{E}_n$, and $\mathcal{H}_n$.

§ 2 An Interpretation of the Language of Second Order Arithmetic

An interpretation of the language of second order arithmetic within a language of division algebras with a distinguished transcendent element is described in this section. The interpretation is based upon the predicates defined in the preceding section. The description of the interpretation is straightforward although somewhat tedious. The reader who is familar with undecidability proofs may wish to omit the details.

Let T be a set of axioms for division rings, for division rings of characteristic 0 or p, or for division algebras over a field k. Let b be a new constant symbol not occurring in the language of T, and let T_b be the theory $T \cup \{\text{Transc}(b)\}$. Each existentially complete model of T can be expanded to a model of T_b by choosing a distinguished transcendental element. Since an existentially complete model of T is finitely homogeneous, all expansions to a model of T_b are isomorphic. The enlarged model is existentially complete for T_b. Conversely, each existentially complete model of T_b is also existentially complete for T. Moreover, a model of T is in one of the classes $\mathcal{F}_T$, $\mathcal{G}_T$, $\mathcal{A}_T$, $\mathcal{H}_n(T)$ if and only if its expansion to model of T_b is in the corresponding class $\mathcal{F}_{T_b}$, $\mathcal{G}_{T_b}$, $\mathcal{A}_{T_b}$, or $\mathcal{H}_n(T_b)$.

Consider now the sentences which are satisfied by an existentially complete model M of T or by its expansion to T_b. The structure M as a model of T satisfies a sentence ϕ in the language of T if and only if M as a model of T_b satisfies ϕ. The following assertions are equivalent for a sentence $\phi(b)$ in the language of T_b: (i) M as a model of T_b satisfies $\phi(b)$; (ii) M as a model of T satisfies $\exists x(\text{Transc}(x) \wedge \phi(x))$; (iii) M as a model of T satisfies $\forall x(\text{Transc}(x) \rightarrow \phi(x))$. Thus, if $\mathcal{C}$ is a class of existentially complete models of T and $\mathcal{C}_b$ is the corresponding class of existentially complete models of T_b, then $\mathcal{TA}(\mathcal{C})$ and $\mathcal{TA}(\mathcal{C}_b)$ are recursively isomorphic.

The interpretation of the language $\mathcal{L}^*$ of second order number theory (see Chapter 7) within the language of T_b proceeds in two stages. The first stage is the elimination of number terms with at least one occurrence of $\cdot$ and another occurrence of either $+$ or $\cdot$. This is necessary because the definition of $\otimes$ in the preceding section required quantifiers. For each number term ξ with at least one occurrence of $\cdot$ and another occurrence of either $+$ or $\cdot$, there is a quantifier-free formula $\phi_\xi(v_0, \ldots, v_n, v_{n+1}, \ldots, v_m)$ such that (i) all variables occurring in ξ have subscript at most n, (ii) ϕ_ξ is a conjunction of formulas of the form $v_i = \xi_j$ where $n + 1 \leq i \leq m$ and ξ_j is a number term either with no occurrences of or else of the form $\xi_{j_1} \cdot \xi_{j_2}$ where ξ_{j_1} and ξ_{j_2} are either constants or variables, (iii) v_m occurs in exactly one conjunct of ϕ_ξ which has the form $v_m = \xi_j$, and (iv)

$$\forall v_0 \ldots \forall v_n \forall v_m (v_m = \xi \leftrightarrow \exists v_{n+1} \ldots \exists v_{m-1}\, \phi_\xi(v_0, \ldots, v_m))$$

and

$$\forall v_0 \ldots \forall v_n \forall v_m (v_m = \xi \leftrightarrow \forall v_{n+1} \ldots \forall v_{m-1}\, \phi_\xi(v_0, \ldots, v_m)).$$

If ξ is a number term which does not have at least one occurrence of $\cdot$ and another occurrence of $+$ or $\cdot$, then let $\phi_\xi(v_0, \ldots, v_n, v_{n+1})$ be the formula $v_{n+1} = \xi$, where all variables occurring in ξ have subscript at most n.

Suppose $\Psi(X_0, \ldots, X_r, v_0, \ldots, v_n)$ is a formula of $\mathcal{L}^*$. The formula $\Psi^\sigma(X_0, \ldots, X_r, v_0, \ldots, v_n)$ is obtained as follows. Let ψ be a particular occurrence of an atomic formula in Ψ. The negation index of ψ is the number of quantifier-free subformulas ϕ of Ψ such that ψ occurs in ϕ and ϕ has the form $\sim\chi$. The action of the transformation σ will depend on the negation index so that the quantifier complexity of Ψ is not increased by σ. (i) Suppose ψ has the form $\xi_1 = \xi_2$ for two number terms ξ_1 and ξ_2. If the negation index of ψ is even, replace ψ by the formula

$$\exists v_{n+1} \ldots \exists v_m \exists v_{m+1} \ldots \exists v_q (\phi_{\xi_1}(v_0, \ldots, v_n, v_{n+1}, \ldots, v_m) \wedge \phi_{\xi_2}(v_0, \ldots, v_n, v_{m+1}, \ldots, v_q) \wedge v_m = v_q).$$

If the negation index of ψ is odd, then replace ψ by the formula

$$\sim \exists v_{n+1} \ldots \exists v_m \exists v_{m+1} \ldots \exists v_q \sim((\phi_{\xi_1}(v_0, \ldots, v_n, v_{n+1}, \ldots, v_m) \wedge \phi_{\xi_2}(v_0, \ldots, v_n, v_{m+1}, \ldots, v_q) \rightarrow v_m =$$

(ii) Suppose ψ has the form $\xi \in X_i$. If the negation index of ψ is even, then replace ψ by the formula

$\exists v_{n+1} \cdots \exists v_m(\phi_\xi(v_0,\ldots, v_n, v_{n+1},\ldots, v_m) \wedge v_m \in X_i)$. If the negation index of ψ is odd, then replace ψ by the formula $\sim\exists v_{n+1} \cdots \exists v_m \sim(\phi_\xi(v_0,\ldots, v_n, v_{n+1},\ldots, v_m) \rightarrow v_m \in X_i)$. Ψ^σ is the formula obtained after application of the above procedure to each atomic formula of Ψ.

The second stage of the interpretation is the transformation of Ψ^σ into a formula in the language of T_b. Let w_i, y_i, z_i, and u_i denote the variables v_{4i}, v_{4i+1}, v_{4i+2}, v_{4i+3} of the language T_b, respectively. One proceeds by induction on the terms ξ and subformulas Φ of Ψ^σ.

(i) ξ is a number term with no occurrences of $\cdot$. Then ξ^ρ is the term obtained by replacing each occurrence of v_i by an occurrence of w_i, each occurrence of a constant s by an occurrence of b^s, and each occurrence of $+$ by an occurrence of $\cdot$ (i.e., the multiplication function in the language of T_b).

(ii) Φ is an atomic formula.

(a) Φ has the form $v_i = \xi$ for a number term ξ. If ξ has no occurrences of $\cdot$, then Φ^ρ is $w_i = \xi^\rho$. Otherwise, ξ has the form $\xi_1 \cdot \xi_2$ where ξ_1 and ξ_2 are either constants or variables. If the negation index of Φ in Ψ^ρ is even, then Φ^ρ is the formula

$$\exists u_i(bu_i = u_i\xi_1{}^\rho \wedge \xi_2{}^\rho u_i = u_i w_i \wedge \sim(u_i = 0)).$$

If the negation index of Φ in Ψ^ρ is odd, then Φ^ρ is the formula

$$\sim\exists u_i \sim((bu_i = u_i\xi_1{}^\rho \wedge \sim(u_i = 0)) \rightarrow \xi_2{}^\rho u_i = u_i w_i).$$

(b) Φ has the form $v_i \in X_j$. Then Φ^ρ is the formula $\text{Subset}(w_i, y_j, z_j)$, where the existential formula for Subset is used if the negation index of Φ in Ψ^ρ is even and the universal formula for Subset is used if the negation index of Φ in Ψ^ρ is odd.

(iii) $(\Phi_1 \vee \Phi_2)^\rho = \Phi_1^\rho \vee \Phi_2^\rho$.

(iv) $(\Phi_1 \wedge \Phi_2)^\rho = \Phi_1^\rho \wedge \Phi_2^\rho$.

(v) $(\sim\Phi)^\rho = \sim\Phi^\rho$.

(vi) $(\exists v_i \Phi(v_i))^\rho = \exists w_i(\text{NNPower}(b, w_i) \wedge (\Phi(w_i))^\rho)$.

(vii) $(\exists X_j \Phi(X_j))^\rho = \exists y_j \exists z_j (\Phi(X_j))^\rho$.

If $\Psi(X_0,\ldots, X_r, v_0,\ldots, v_n)$ is a formula of $\mathcal{L}^*$, then let $\Psi^\tau(y_0, z_0,\ldots, y_r, z_r, w_0,\ldots, w_n) = (\Psi^\sigma)^\rho$. The formula Ψ^τ is a formula in the language of T_b in which each constant s has been replaced by b^s, number variables have been relativized to the nonnegative powers of b, and set variables have been relativized to subsets of the nonnegative powers of b.

Proposition 16.4. Suppose that D is an existentially complete structure for T_b and that $\mathfrak{N}_D$ is the structure for second order arithmetic determined by D. Let $A_0, A_1,\ldots$ be a sequence of subsets of N in $\mathfrak{N}_D$ and let $(c_0, d_0), (c_1, d_1), \ldots$ be a sequence of pairs of elements of D for which $A_j = \{n :$ D satisfies $\text{Subset}(b^n, c_j, d$ For each formula $\Psi(X_0,\ldots, X_r, v_0,\ldots, v_m)$ of $\mathcal{L}^*$, $\mathfrak{N}_D$ satisfies $\Psi(A_0,\ldots, A_r, n_0,\ldots, n_m)$ if and only if D satisfies $\Psi^\tau(c_0, d_0, c_1, d_1,\ldots, c_r, d_r, b^{n_0},\ldots, b^{n_m})$.

Proof. This is a consequence of the definition of $\mathfrak{N}_D$ and the definition of the interpretation τ.

One should observe that if Ψ is a formula consisting of a sequence of quantifiers followed by a quantifier-free formula, then Ψ is logically equivalent to a formula in prenex normal form with at most one more alternation of quantifiers than the formula Ψ.

§ 3 Second Order Arithmetic and Classes of Existentially Complete Division Algebras

Properties of the structures $\mathfrak{N}_D$ will be used to compute the degrees of unsolvability of the theories of particular classes of existentially complete division algebras. Assume for the remainder of this chapter that T is a countable theory and that the set of Gödel numbers of sentences in T is an arithmetical set.

Proposition 16.5. Assume D is an existentially complete model of T_b. If $\Psi(v_0,\ldots, v_m)$ is a formula of first order arithmetic and $n_0, \ldots, n_m$ are nonnegative integers, then N satisfies $\Psi(n_0,\ldots, n_m)$ if and only if D satisfies $\Psi^{\tau}(b^{n_0},\ldots, b^{n_m})$.

Proof. This is a consequence of the preceding proposition and the fact that $\{b^n : n \in N\}$ with the operations of $\oplus$ and $\otimes$ is isomorphic to N.

Corollary 16.6. The theory $\mathcal{T}h(N)$ is one-one reducible to the theory $\mathcal{T}h(\mathcal{E}_{T_b})$ and consequently is one-one reducible to the theory $\mathcal{T}h(\mathcal{E}_T)$.

Proof. Each formula Ψ in $\mathcal{T}h(N)$ has no free variables. According to the preceding proposition, N satisfies Ψ if and only if each existentially complete model of T_b satisfies Ψ^{τ}. Therefore, Ψ is in $\mathcal{T}h(N)$ if and only if Ψ^{τ} is in $\mathcal{T}h(\mathcal{E}_{T_b})$.

Theorem 16.7. The theory $\mathcal{T}h(N)$ is one-one reducible to $T_b{}^f$ and consequently is one-one reducible to T^f. Hence, T^f and $\mathcal{T}h(N)$ are recursively isomorphic.

Proof. The same argument as in the proof of Corollary 16.6 shows that $\mathcal{T}h(N)$ is one-one reducible to T^f. On the other hand, Theorem 7.7 asserts that T^f is one-one reducible to $\mathcal{T}h(N)$. Therefore, T^f

and $\mathcal{Th}(N)$ are recursively isomorphic (see H. Rogers (87)).

The computation of the degree of unsolvability of T^F depends upon the following lemma.

Lemma 16.8. If A is a set of natural numbers and D is an existentially universal model of T_b, then there are elements c and d of M such that a nonnegative integer n is an element of A if and only if D satisfies Subset(b^n, c, d).

Proof. A division algebra generated by two elements, say s and in which $\{s^{-n}ts^n : n \in N\}$ was a set of commuting, algebraically independent (over the center) elements, was constructed in the proof of Theorem 14.30. Moreover, the element s could be required to be transcendental over the center. The model D must include such a division subalgebra, since D is existentially universal. Since D is finitely homogeneous, one may take s to be the element b of D. Let the element c of D be an appropriate t. Consider the subfield C generated by $\{c_n = b^{-n}cb^n : n < \omega\}$. Let f be the automorphism of C defined by

$$f(c_n) = \begin{cases} c_n & \text{if } n \notin A \\ c_n^{-1} & \text{if } n \in A \end{cases}$$

According to Theorem 14.1, there is a extension D' of D and an element d' of D' for which $(d')^{-1}c_n d' = f(c_n)$ for all $n \in N$. The division subalgebra generated by b, c, and d' in D' is completely described by the existential formulas satisfied by b, c, an d' in D', that is, by the existential type in D' of d' over b and c. Since D is existentially universal, there is an element d of D which realizes the same existential type in D. Then $d^{-1}c_n d = f$ for all n, so $c_n d \neq dc_n$ if and only if n is in A. Thus n is in A if and only if D satisfies Subset(b^n,c,d).

Proposition 16.9. Assume D is an existentially universal model of T_b. If $\{A_i : i < \omega\}$ is a sequence of subsets of N and $\{n_i : i < \omega\}$ is a sequence of elements of N, then there are sequences $\{c_i : i < \omega\}$ and $\{d_i : i < \omega\}$ of elements of D such that for each formula $\Psi(X_0,\ldots, X_r, v_0,\ldots, v_m)$ of $\mathcal{L}^*$, $\mathfrak{N}$ satisfies $\Psi(A_0,\ldots, A_r, n_0,\ldots, n_m)$ if and only if D satisfies $\Psi^{\tau}(c_0, d_0, c_1, d_1,\ldots, c_r, d_r, b^{n_0},\ldots, b^{n_m})$.

Proof. This is a consequence of Lemma 16.8 and Proposition 16.4.

Corollary 16.10. Assume D is an existentially universal model of T_b. Then $\mathfrak{N}$ satisfies a sentence Ψ of $\mathcal{L}^*$ if and only if D satisfies Ψ^{τ}.

Proposition 16.11. If D is an infinitely generic model of T_b, then $\mathfrak{N}_D \prec_{\mathcal{L}^*} \mathfrak{N}$.

Proof. This result is a consequence of Proposition 16.9, Proposition 16.4, and the fact that each infinitely generic model is an elementary substructure of an existentially universal model.

Theorem 16.12. Second order number theory is one-one reducible to the theory T_b^F and consequently is one-one reducible to T^F. Hence, T^F and $\mathfrak{Th}_2(\mathfrak{N})$ are recursively isomorphic.

Proof. A sentence Ψ of $\mathcal{L}^*$ is true in $\mathfrak{N}$ if and only if it is true in $\mathfrak{N}_D$ for each infinitely generic model D (Proposition 16.11). Therefore, Ψ is true in $\mathfrak{N}$ if and only if Ψ^{τ} is in T_b^F, so the theory of second order arithmetic is one-one reducible to T_b^F. Since T_b^F and T^F are recursively isomorphic, the theory of second order arithmetic is one-one reducible to T^F. On the other hand, T^F is one-one reducible to the theory of second order arithmetic (Theorem 7.6). Hence, T^F and $\mathfrak{Th}_2(\mathfrak{N})$ are recursively isomorphic.

Proposition 16.13. The chain $\{\mathcal{H}_n : n < \omega\}$ is strictly descending.

Proof. Otherwise, for some integer n, $\mathcal{H}_n = \mathcal{H}_{n+1} = \cdots = \mathcal{G}_T$. But then $\mathcal{T}\mathcal{A}(\mathcal{G}_T)$ would be a Π^1_{2n-1} set, which would contradict the preceding theorem.

Proposition 16.14. The diagram of an infinitely generic model of cannot be an analytical set.

Proof. This is a consequence of Theorem 16.12 and Proposition 7.1:

Theorem 16.15. $T^F \neq T^f$.

Proof. The theory T^f is a hyperarithmetical set whereas T^F is not even an analytical set.

Theorem 16.16. There are $2^{\aleph_0}$, pairwise non-elementarily equival existentially complete models of T.

Proof. This is a consequence of Theorem 16.12 and Corollary 7.18.

Theorem 16.16 was proven originally by A. Macintyre (58) for T_0 and T_p, $p \neq 2$.

Further investigation of the theories T^F and T^f as well as of the classes $\mathcal{E}_n$ and $\mathcal{H}_n$ depends upon the $\mathcal{E}_{T_b}$-persistent formulas. Assume that $\Psi(X_0, \ldots, X_r)$, a formula of $\mathcal{L}^*$, has no second order quantifiers and no free occurrences of number variables. Suppose that M and M' are existentially complete models of T_b and M' extends M. Let $c_0, d_0, \ldots, c_r, d_r$ be elements of M. Since Ψ has neither second order quantifiers nor free number variables, the formula $\Psi^\tau(c_0, d_0, \ldots, c_r, d_r)$ asserts the truth of some arithmetical relationship (i.e., number quantifiers only) involving certain subsets of $\{b^n : n < \omega\}$, some of which may be defined by the pairs (c_j, d_j). For each j, the set $\{b^n : (b^{-n} c_j b^n)d_j \neq d_j(b^{-n} c_j b^n)\}$ is certain the same set in both M and M'. Therefore, $\Psi^\tau(c_0, d_0, \ldots, c_r, d_r)$

holds in M if and only if it holds M'. In other words, the formula $\Psi^{\tau}(y_0, z_0, \ldots, y_r, z_r)$ is persistent under both extension and restriction in the class $\mathcal{E}_{T_b}$. Consequently, the formula

$$\exists y_{j_1} \exists z_{j_1} \ldots \exists y_{j_s} \exists z_{j_s} \Psi^{\tau}(y_0, z_0, \ldots, y_r, z_r)$$

is $\mathcal{E}_{T_b}$-persistent also (see Chapter 4).

Define classes S_n and R_n of formulas of $\mathcal{L}^*$ by induction as follows. Let $S_0 = R_0 = \{\Psi(X_0, \ldots, X_r) : \Psi$ contains neither second order quantifiers nor free occurrences of number variables$\}$. Suppose S_n and R_n have been defined. Let $S_{n+1} = S_n \cup R_n \cup \{\exists X_{j_1} \ldots \exists X_{j_s} \Psi(X_0, \ldots, X_r) : \Psi(X_0, \ldots, X_r)$ is a formula in $R_n\}$, and let $R_{n+1} = \{\sim\Psi : \Psi$ is in $S_{n+1}\}$. Each formula $\Psi^{\tau}(y_0, z_0, \ldots, y_r, z_r)$ obtained from a formula $\Psi(X_0, \ldots, X_r)$ in S_1 is $\mathcal{E}_{T_b}$-persistent.

Let the set F'_{n+1} of formulas be $\{\Psi^{\tau}(y_0, z_0, \ldots, y_r, z_r) : \Psi$ is in $S_n\} \cup \{\phi : \phi$ is an $\exists_{n+2}$ formula in the langauge of $T_b\}$. The classes $\mathcal{E}_n(T_b)$ and $\mathcal{H}_n(T_b)$ are as defined in Chapter 4.

Let the set F_{n+1} of formulas be $\{\mathrm{Transc}(x) \wedge \Psi^{\tau}(y_0, z_0, \ldots, y_r, z_r, x) : \Psi(X_0, \ldots, X_r)$ is in S_n and $\Psi^{\tau}(y_0, z_0, \ldots, y_r, z_r) = \Psi^{\tau}(y_0, z_0, \ldots, y_r, z_r, b)\} \cup \{\phi : \phi$ is an $\exists_{n+2}$ formula in the language of $T\}$. The subclasses $\mathcal{E}_n(T)$ of Σ_T are defined relative to the sets F_n of formulas as in Chapter 4. Clearly, an existentially complete model of T is in $\mathcal{E}_n(T)$ if and only if its expansion to a model of T_b is in $\mathcal{E}_n(T_b)$. The theories of $\mathcal{E}_n(T)$ and $\mathcal{E}_n(T_b)$ are related in the same way as the theories of $\mathcal{E}_T$ and $\mathcal{E}_{T_b}$.

As corresponding classes of existentially complete models of T and T_b contain precisely the same division algebras, the distinction between corresponding classes will not be observed rigorously, espccially in statements of results. The classes relative to T_b are used in proofs for the sake of convenience. For example, the following theorem is true for both $\mathcal{E}_n(T)$ and $\mathcal{E}_n(T_b)$. The proof is given only for $\mathcal{E}_n(T_b)$, but the result holds for $\mathcal{E}_n(T)$ because $\mathcal{Th}(\mathcal{E}_n(T))$ and $\mathcal{Th}(\mathcal{E}_n(T_b))$ are recursively isomorphic.

Theorem 16.17. (i) The theory $\mathcal{Th}(\mathcal{E}_n(T))$ is a complete Π^1_n s[et] for $n \geq 1$.

(ii) The theory $\mathcal{Th}(\mathcal{H}_n(T))$ is a complete Π^1_{2n-1} set for $n \geq 1$.

Proof. (i) Let A_n be the canonical complete Π^1_n set (see H. Rogers (87)). There is a Π^1_n formula $\Psi(x)$, in $\mathcal{L}^*$ prenex normal form (see Chapter 7), with one free number variable x, such that for each number m in N, m is in A if and only if $\mathcal{N}$ satisfies $\Psi(m)$. According to Corollary 16.10, $\mathcal{N}$ satisfies $\Psi(m)$ i[f] and only if each existentially universal model of T_b satisfies $\Psi^\tau(b^m)$.

Suppose m is in A. The formula $\Psi(m)$ is in the class R_n, so $\Psi^\tau(b^m)$ is persistent under restriction in the class $\mathcal{E}_n$. Since each member of $\mathcal{E}_n(T_b)$ is included in an existentially universal model of T_b which satisfies $\Psi^\tau(b^m)$, each member of $\mathcal{E}_n(T_b)$ satisfies $\Psi^\tau(b^m)$. Therefore, $\Psi^\tau(b^m)$ is in $\mathcal{Th}(\mathcal{E}_n(T_b))$.

Conversely, suppose m is not in A. Let D be an existentially universal model of T_b. The model D is in the class $\mathcal{E}_n(T_b)$, and D does not satisfy the sentence $\Psi^\tau(b^m)$. Therefore, $\Psi^\tau(b^m)$ is not in $\mathcal{Th}(\mathcal{E}_n(T_b))$.

Thus, m is in A_n if and only if the sentence $\Psi^\tau(b^m)$ is in $\mathcal{Th}(\mathcal{E}_n(T_b))$. This shows that A_n in one-one reducible to $\mathcal{Th}(\mathcal{E}_n($[…] On the other hand, $\mathcal{Th}(\mathcal{E}_n(T_b))$ is Π^1_n set (Lemma 7.4), so $\mathcal{Th}(\mathcal{E}_n(T_b))$ is one-one reducible to A_n. Hence, A_n and $\mathcal{Th}(\mathcal{E}_n(T_b))$ are recursively isomorphic. Since $\mathcal{Th}(\mathcal{E}_n(T_b))$ and $\mathcal{Th}(\mathcal{E}_n(T))$ are recursively isomorphic, $\mathcal{Th}(\mathcal{E}_n(T))$ is a complete Π^1_n set.

(ii) The proof is similar to that of part (i).

Corollary 16.18. $\mathcal{TA}(\mathcal{E}_T)$ is not a hyperarithmetical set.

One may observe that Corollary 16.18 and Theorem 7.17 yield another proof of Theorem 16.16.

The proof of Theorem 16.17 also provides information concerning sentences in T^f whose negations are in T^F. The canonical complete Π^1_1 set A_1 cannot be one-one reducible to $T_b{}^f$, because $T_b{}^f$ is hyperarithmetical. Yet $\Psi^\tau(b^m)$ is in $\mathcal{TA}(\mathcal{E}_{T_b})$ and so is in $T_b{}^f$ for each integer m in A_1. Therefore, it must be that there are nonnegative integers m for which $\Psi^\tau(b^m)$ is in $T_b{}^f$ although m is not in A_1. Indeed, there must be infinitely many such integers, for otherwise a one-one reduction could be obtained by excluding finitely many integers. Moreover, it must be impossible to deduce, from finitely many instances of m in $N - A_1$ but $\Psi^\tau(b^m)$ in $T_b{}^f$, for which other integers m' is it true that m' is in $N - A_1$ but $\sim\Psi^\tau(b^{m'})$ is not in $T_b{}^f$. Thus, there are $\aleph_0$ many "independent" sentences $\Psi^\tau(b^m)$ for which $\Psi^\tau(b^m)$ is in $T_b{}^f$ and $\sim\Psi^\tau(b^m)$ is in $T_b{}^F$. The formula $\Psi^\tau(x)$ is equivalent (after a suitable change of subscripted variables if necessary) to an $\forall_6$ formula $\Phi(x)$. For each m such that $\Psi^\tau(b^m)$ is in $T_b{}^f$ and m is not in A_1, the $\forall_6$ sentence $\forall x(\mathrm{Transc}(x) \to \Phi(x^m))$ is in T^f and its negation, the $\exists_6$ sentence $\exists x(\mathrm{Transc}(x) \wedge \sim\Phi(x^m))$ is in T^F. Thus, there are $\aleph_0$ many "independent" $\forall_6$ sentences in T^f whose negations are in T^F.

A stronger result is true for the theory of groups. A. Macintyre (56) has exhibited an $\forall_4$ sentence, in the langauge of groups, which is in the finite forcing companion of the theory of groups and whose negation is in the infinite forcing companion of the theory of groups. A subtler method than that of the preceding paragraph will produce $\forall_4$ sentences for the case of division algebras over fields with recursive diagrams.

Let A be a recursively enumerable, nonrecursive set of nonnegative integers. There is an existential formula $S_A(x)$ of first order arithmetic such that each nonnegative integer m is in A if and only if N satisfies $S_A(m)$ (see Matijasevic (65,66)). Let Φ_A be the comprehension axiom for A, that is, the formula $\exists X \forall x(x \in X \leftrightarrow S_A(x))$. The formula Φ_A is logically equivalent to a formula of $\mathcal{L}^*$ with three alternations of quantifiers beginning with an existential quantifier. Therefore, Φ_A^{τ} is logically equivalent to an $\exists_4$ formula; for convenience, denote this formula by Φ_A^{τ} also. The sentence Φ_A^{τ} is true in each existentially universal model of T_b, because it is true in $\mathcal{N}$, so Φ_A^{τ} is in T_b^{F}. It remains to show that $\sim\Phi_A^{\tau}$ is in T_b^{f}.

Proposition 16.19 . Assume T is a recursively axiomatized theory. There is a finitely generic model of T_b which satisfies $\sim\Phi_A^{\tau}$.

Proof. This is a special case of a type omitting theorem of A. Macintyre (57). For the sake of completeness, a proof of Proposition 16.19 is included.

Form a new language $\mathcal{L}(E)$ by augmenting the language of T_b with an infinite set $E = \{e_i : i < \omega\}$ of new constants. Let $\{\zeta_n = (c_n, d_n) : n < \omega\}$ be an enumeration of all ordered pairs of constants in $\mathcal{L}(E)$ and let $\{\phi_n : n < \omega\}$ be an enumeration of all sentences of $\mathcal{L}(E)$. Construct a complete sequence of conditions in $\mathcal{L}(E)$ relative to T_b as follows.

Step 0. Let P_0 be the condition $\{be_0 = 1\}$.

Step 2n + 1. If $P_{2n} \Vdash \phi_n$ or $P_{2n} \Vdash \sim\phi_n$, then let $P_{2n+1} = P_n$. Otherwise, there is a condition $Q \supseteq P_{2n}$ such that $Q \Vdash \phi_n$. Let $P_{2n+1} = Q$.

Step 2n + 2. This step will ensure that the elements c_n and d_n of the order pair ζ_n do not determine the set $\{b^m : m \in A\}$. If $T_b \cup P_{2n+1}$ implies either $c_n = 0$ or $d_n = 0$, then let $P_{2n+2} = P_{2n+1}$. Otherwise it is consistent to assume $c_n \neq 0$,

$d_n \neq 0$. Since A is not recursive, the set of formulas

$$\{e_0^m c_n b^m d_n = d_n e_0^m c_n b^m : m \notin A\} \cup \{\sim(e_0^m c_n b^m d_n = d_n e_0^m c_n b^m) : m \in A\}$$

cannot be recursive in $T_b \cup P_{2n+1} \cup \{\sim(c_n = 0), \sim(d_n = 0)\} = \overline{T}_b$. Therefore, there is a natural number m for which either (i) m is in A and $\overline{T}_b \cup \{e_0^m c_n b^m d_n = d_n e_0^m c_n b^m\}$ is consistent or (ii) m is not in A and $\overline{T}_b \cup \{\sim(e_0^m c_n b^m d_n = d_n e_0^m c_n b^m)\}$ is consistent. If alternative (i) is true, then let $P_{2n+2} = P_{2n+1} \cup \{\sim(c_n = 0), \sim(d_n = 0), e_0^m c_n b^m d_n = d_n e_0^m c_n b^m\}$. Otherwise alternative (ii) is true, and let $P_{2n+2} = P_{2n+1} \cup \{\sim(c_n = 0), \sim(d_n = 0), \sim(e_0^m c_n b^m d_n = d_n e_0^m c_n b^m)\}$.

This completes the construction.

Let $S = \{P_n : n < \omega\}$ and let D_S be the finitely generic structure determined by the complete sequence S. Suppose c and d are nonzero elements of D_S. Then $(c, d) = \zeta_n$ for some n, and since D_S is a model of P_{2n+2},

$$\{b^m : D_S \text{ satisfies } \mathrm{Subset}(b^m, c, d)\} \neq \{m : m \in A\}.$$

Therefore, D_S satisfies $\sim\Phi_A^{\tau}$.

<u>Theorem 16.20</u>. Assume T is a recursively axiomatized theory. There is an $\exists_4$ sentence in T^F whose negation is in T^f.

<u>Proof</u>. Let ϕ be the sentence $\exists x(\mathrm{Transc}(x) \wedge \Phi_A^{\tau}(x))$, where $\Phi_A^{\tau} = \Phi_A^{\tau}(b)$. First, suppose all models of T have the same characteristic. Then T has the joint embedding property, so T^f is complete. Since Φ_A is not in T_b^f, ϕ is not in T^f. Therefore $\sim\phi$ is in T^f. On the other hand, Φ_A^{τ} is in T_b^F, so ϕ is in T^F.

Now suppose T has models of different characteristic. Since ϕ is independent of characteristic, ϕ holds in all infinitely generic models of T and $\sim\phi$ holds in all finitely generic models of T. Thus, ϕ is in T^F and $\sim\phi$ is in T^f.

Suppose A is an arithmetical set of natural numbers which is not recursive in T. Let ϕ_A be the comprehension axiom for A. An argument similar to the preceding one proves that the formula $\exists x(\mathrm{Transc}(x) \wedge \Phi_A{}^{\tau}(x))$ is in T^F and its negation is in T^f.

The comprehension axioms for analytical sets can be used to refine the result that no infinitely generic model of T has an analytical diagram. Let A be an analytical set. If A is a Δ_1^1 set or a Δ_n^1 set but neither a Σ_{n-1}^1 set nor a Π_{n-1}^1 set $(n > 1)$, then let $P_A(x)$ and $Q_A(x)$ be Σ_n^1 and Π_n^1 formulas defining A, respectively. The comprehension axiom Φ_A for A is the sentence

$$\exists X \forall x(x \in X \rightarrow P_A(x) \wedge Q_A(x) \rightarrow x \in X).$$

If A is a Σ_n^1 set but not a Π_n^1 set (a Π_n^1 set but not a Σ_n^1 set), then let $P_A(x)$ be a Σ_n^1 formula (Π_n^1 formula) defining A. The comprehension axiom Φ_A for A is the sentence

$$\exists X \ \forall x(x \in X \leftrightarrow P_A(x)).$$

If A is a Δ_n^1 set, then Φ_A is a Σ_n^1 formula. If A is Σ_n^1 but not Π_n^1 or Π_n^1 but not Σ_n^1, then Φ_A is a Σ_{n+1}^1 formula.

Note that each existentially universal model D must satisfy $\Phi_A{}^{\tau}$ because $\mathfrak{N}_D = \mathfrak{N}$.

Lemma 16.21. If A is a Δ_n^1 set of natural numbers and D is a model in $\mathcal{E}_{n+1}(T_b)$, then there are elements c and d of D such that, for each number m, m is in A if and only if D satisfies $\mathrm{Subset}(b^m, c, d)$.

Proof. The sentence Φ_A is a Σ_n^1 sentence, since A is a Δ_n^1 set, and so Φ_A is in S_n and $\Phi_A{}^{\tau}$ is in F'_{n+1}. The sentence $\Phi_A{}^{\tau}$ is true in each existentially universal model. Since each member of $\mathcal{E}_{n+1}(T_b)$ is F_{n+1}-persistently complete and is a substruct of an existentially universal model, $\Phi_A{}^{\tau}$ is in $\mathcal{TA}(\mathcal{E}_{n+1}(T_b))$. Therefore, there are elements c and d of D for which D satisfie $\Psi_A{}^{\tau}(c, d)$, where $\Psi_A(X)$ is either the formula

$\forall x(x \in X \to P_A(x) \wedge Q_A(x) \to x \in X)$ or the formula $\forall x(x \in X \leftrightarrow P_A(X))$, whichever is appropriate. The formula $\Psi_A(X)$ is a Σ^1_n formula also, so $\Psi_A{}^\tau(y, z)$ is $\mathcal{E}_{n+1}$-persistent. Consequently, if D' is an existentially universal model extending D, then D' satisfies $\Psi_A{}^\tau(c, d)$ also. Then, for each natural number m, m is in A if and only if D' satisfies Subset(b^m, c, d), because $\mathfrak{n}_{D'} = \mathfrak{n}$. But D' satisfies Subset(b^m, c, d) if and only if D satisfies Subset(b^m, c, d). Hence, for each natural number m, m is in A if and only if D satisfies Subset(b^m, c, d).

Proposition 16.22. If D is a countable member of $\mathcal{E}_n(T)$ for an $n \geq 2$, then the diagram of D is at least a Δ^1_{n-1} set, that is, the diagram of D is neither a Σ^1_{n-2} set nor a Π^1_{n-2} set.

Proof. The expansion of D to a model of T_b is in $\mathcal{E}_n(T_b)$. Let b^{-1} denote the name in the diagram of D for the multiplicative inverse of b. Let A be a subset of N which is Δ^1_{n-1} but neither Σ^1_{n-2} nor Π^1_{n-2}. Then the same is true of N - A. According to the preceding lemma, there are elements c and d of D such that m is in A, that is, m is not in N - A, if and only if D satisfies ~Subset(b^m, c, d). In other words, m is in A if and only if the formula $(b^{-1})^m \, c \, b^m \, d = d \, (b^{-1})^m \, c \, b^m$ is in the diagram of D. Therefore, A is one-one reducible to the positive diagram of D, which is recursive in the diagram of D. Hence, the diagram of D can be neither Σ^1_{n-2} nor Π^1_{n-2}.

Proposition 16.23. If D is a member of $\mathcal{H}_n(T)$ for an $n \geq 2$, then the diagram of D is at least a Δ^1_{2n-2} set, that is, the diagram of D is neither a Σ^1_{2n-3} set nor a Π^1_{2n-3} set.

Proof. The proof is similar to that of Proposition 16.22.

The analogues for groups of the results in this chapter are true also. The arithmetical definition of structures for second order arithmetic is easier in the case of groups, because the subgroup generated by an element consists of just the powers of that element. Consequently, one may use the formula $\forall z(xz = zx \rightarrow yz = zy)$ for Power(x, y). Also, conjuncts which assert that an element is nonzero should be omitted. Otherwise, the adaptation of section one to groups is straightforward. Section two is unchanged for groups. Finally, the appropriate analogues for groups of all results in section three are true.

BIBLIOGRAPHY

1. Amitsur, S.A. "A Generalization of Hilbert's Nullstellensatz." Proceedings of the American Mathematical Society, vol.8 (1957), pp. 649-656.

2. __________. "Generalized Polynomial Identities and Pivotal Monomials." Transactions of the American Mathematical Society, vol.114 (1965), pp.210-226.

3. __________. "Rational Identities and Applications to Algebra and Geometry." Journal of Algebra, vol.3 (1966), pp.304-359.

4. Ax, J. and S. Kochen. "Diophantine Problems over Local Fields, III. Decidable Fields." Annals of Mathematics, vol.83 (1966), pp. 437-456.

5. Bacsich, P.D. "Defining Algebraic Elements." Journal of Symbolic Logic, vol.38 (1973), pp.93-101.

6. Barwise, J., and A. Robinson. "Completing Theories by Forcing." Annals of Mathematical Logic, vol.2 (1970), pp.119-142.

7. Bell, J.L. and A.B. Slomson. Models and Ultraproducts. Amsterdam, North-Holland Publishing Company, 1969.

8. Boffa, M. "Modèles universels homogenes et modèles génériques." C.R. Acad. Sc. Paris, vol.274 (1972), pp.693-694.

9. __________. "Corps λ-clos." C.R. Acad. Sc. Paris, vol.275 (1972), pp.881-882.

0. __________. "Sur l'existence des corps universals-homogenes." C.R. Acad. Sc. Paris, vol.275 (1972), pp.1267-1268.

1. Boffa, M., and P. van Praag. "Sur les corps génériques." C.R. Acad. Sc. Paris, vol.274 (1972), pp.1325-1327.

2. __________. "Sur les sous-champs maximaux des corps génériques dénombrables." C.R. Acad. Sc. Paris, vol.275 (1972), pp.945-948.

3. Carson, A.B. "The Model-Completion of the Theory of Commutative Regular Rings." Journal of Algebra, vol.27 (1973), pp.136-146.

4. Cherlin, G. A New Approach to the Theory of Infinitely Generic Structures. Dissertation, Yale University, 1971.

5. __________. "The Model-Companion of a Class of Structures." Journal of Symbolic Logic, vol.37 (1972), pp.546-556.

6. __________. "Second Order Forcing, Algebraically Closed Structures, and Large Cardinals." Preprint.

7. __________. "Algebraically Closed Commutative Rings." Journal of Symbolic Logic, vol.38 (1973), pp.493-499.

18. Cohen, P.J. Set Theory and the Continuum Hypothesis. New York, W.A. Benjamin, Inc., 1966.

19. Cohn, P.M. "Free Ideal Rings." Journal of Algebra, vol.1 (1964), pp.47-69.

20. __________. Universal Algebra. New York, Harper and Row, 1965.

21. __________. "The Embedding of Firs in Skew Fields." Proceedings of the London Mathematical Society, vol.23 (1971), pp.193-213.

22. __________. "Un critère d'immersibilité d'un anneau un corps gauche." C.R. Acad. Sc. Paris, vol.272 (1971), pp.1442-1444.

23. __________. "Rings of Fractions." American Mathematical Monthly, vol.78 (1971), pp.596-615.

24. __________. "Universal Skew Fields of Fractions." Symposia Mathematica, vol.8 (1972), pp.135-148.

25. __________. "Skew Fields of Fractions, and the Prime Spectrum of a General Ring." Tulane Symposium on Rings and Operator Algebras, 1970-1971, vol.I. Lecture Notes in Mathematics, vol.246. Berlin, Springer-Verlag, 1972.

26. __________. "Rings of Fractions." Lectures at the University of Alberta, 1972.

27. __________. Free Rings and Their Relations. London, Academic Press, 1971.

28. __________. "Presentations of Skew Fields, I." Preprint.*

29. Cusin, R. "Sur le "forcing" en theorié des modèles." C.R. Acad. Sc. Paris, vol.272 (1971), pp.845-848.

30. __________. "Structures prehomogenes et structures génériques." C.R. Acad. Sc. Paris, vol.273 (1971), pp.137-140.

31. __________. "Recherche du "forcing-compagnon" et du "modèle-compagnon" d'une theorié, liée a l'existence des mòdeles $\aleph_\alpha$-universeles." C.R. Acad. Sc. Paris, vol.773 (1971), pp.956-959.

32. __________. "L'Algèbre des formules d'une theorié complete et 'forcing-compagnon'." C.R. Acad. Sc. Paris, vol.275 (1972), pp.1269-1272.

33. Cusin, R., and J.R. Pabion. "Structures générique associees à une classe de théories." C.R. Acad. Sc. Paris, vol.172 (1971), pp.1620-1623.

34. Eklof, P., and G. Sabbagh. "Model-Completions and Modules." Annals of Mathematical Logic, vol.2 (1970), pp.251-295.

35. Faith, C.C. "On Conjugates in Division Rings." Canadian Journal of Mathematics, vol.10 (1958), pp.374-380.

36. Feferman, S., D. Scott, and S. Tennenbaum. "Models of arithmetic through function rings." Notices of the American Mathematical Society, vol.173 (1959), 556-31.

* Proceedings of the Cambridge Philosophical Society, vol.77 (1975).

37. Fisher, E., and A. Robinson. "Inductive Theories and Their Forcing Companions." Israel Journal of Mathematics, vol.12 (1972), pp.95-107.

38. Friedman, H. "Countable Models of Set Theory." To appear in the proceedings of the symposium in Cambridge, 1971.

39. Gaifman, H.A. "A note on models and submodels of arithmetic." Conference on Mathematical Logic, London, 1970. Lecture Notes in Mathematics, vol.255. Berlin, Springer-Verlag, 1972.

40. Grezegorzyk, A., A. Mostowski, and C. Ryll-Nardzewski. "The classical and the ω-complete arithmetic." Journal of Symbolic Logic, vol.23 (1956), pp.188-206.

41. Henrard, P. "Forcing with Infinite Conditions." Conference on Mathematical Logic, London, 1970. Lecture Notes in Mathematics, vol.255. Berlin, Springer-Verlag, 1972.

42. __________. "Une theorié sans modèle générique." C.R. Acad. Sc. Paris, vol.272 (1971), pp.293-294.

43. __________. "Le "Forcing Companion" sans "forcing"." C.R. Acad. Sc. Paris, vol.276 (1973), pp.821-822.

44. Herstein, I.N. "Conjugates in Division Rings." Proceedings of the American Mathematical Society, vol.7 (1956), pp.1021-1022.

45. Higman, C., B.H. Newmann, and H. Newmann. "Embedding Theorems for Groups." Journal of the London Mathematical Society, vol.24 (1949), pp.247-254.

46. Hirschfeld, J. Existentially Complete and Generic Structures in Arithmetic. Dissertation, Yale University, 1972.

47. Jacobson, N. "A Note on Non-Commutative Polynomials." Annals of Mathematics (2), vol.35 (1934), pp.209-210.

48. __________. Structure of Rings. Providence, Rhode Island, American Mathematical Society, 1968.

49. Jonsson, B. "Algebraic Extensions of Relational Systems." Mathematica Scandinavica, vol.11 (1962), pp.179-205.

50. Karp, C. Languages with Expressions of Infinite Length. Amsterdam, North-Holland Publishing Company, 1964.

51. __________. "Finite Quantifier Equivalence." The Theory of Models, Proceedings of the 1963 Symposium at Berkeley. Amsterdam, North-Holland Publishing Company, 1965.

52. Keisler, H.J. Model Theory for Infinitary Logic. Amsterdam, North-Holland Publishing Company, 1971.

53. __________. "Forcing and the Omitting Types Theorem." Preprint.

54. Kueker, D.W. "Definability, Automorphisms, and Infinitary Languages." Syntax and Semantics of Infinitary Languages. Lecture Notes in Mathematics, vol.72 , Berlin, Springer-Verlag, 1968.

55. Lerman, M. "Recursive functions modulo co-maximal rings." Transactions of the American Mathematical Society, vol.148 (1970), pp.429-444.

56. Macintyre, A. "On Algebraically Closed Groups." Annals of Mathematics, vol.96 (1972), pp.53-97.

57. ________. "Omitting Quantifier-Free Types in Generic Structures Journal of Symbolic Logic, vol.37 (1972), pp.512-520.

58. ________. "On Algebraically Closed Division Rings." Preprint.

59. ________. "Martin's Axiom Applied to Existentially Closed Groups." Mathematica Scandinavica, vol.32 (1973), pp.46-56.

60. ________. "The Word Problem for Division Rings." Journal of Symbolic Logic, vol.38 (1973), pp.428-435.

61. ________. "A note on Axioms for Infinite Generic Structures." To appear.

62. ________. "Model-Completeness for Sheaves of Structures." Fundamenta Mathematicae, vol.81 (1973), pp.73-89.

63. Macintyre, A., H. Simmons, and A. Goldrei. "The Forcing Companions of Number Theories." Israel Journal of Mathematics, vol.14 (1973), pp.317-337.

64. Malcev, A. "On the Immersion of an Algebraic Ring into a Field." Mathematishe Annalen, vol.113 (1937), pp.686-691.

65. Matijasevič, Y.V. "Diophantive Representation of Enumerable Predicates." Dokl. Akad. Nauk. S.S.S.R., vol. 191 (1970), pp.279-282. Translated by F.M. Goldware, Mathematics of the U.S.S.R., Izvestiza, vol.15 (1971), pp.1-28.

66. ________. "Diophantine Representation of Recursively Enumerable Predicates." Proceedings of the Second Scandinavian Logic Symposium, Oslo, 1970.

67. Mostowski, A., and Y. Suzuki. "On ω-models which are not β-models. Fundamenta Mathematicae, vol.65 (1969), pp.83-93.

68. Neumann, B.H. "Some Remarks on Infinite Groups." Journal of the London Mathematical Society, vol.12 (1937), pp.120-127.

69. ________. "Adjunction of Elements to Groups." Journal of the London Mathematical Society, vol.18 (1943), pp.4-11.

70. ________. "A Note on Algebraically Closed Groups." Journal of the London Mathematical Society, vol.27 (1952), pp.247-249.

71. ________. "The Isomorphism Problem for Algebraically Closed Groups." Word Problems (edited by W.W. Boone et al). Amsterdam, North-Holland Publishing Company, 1973.

72. Ore, Oystein. "Theory of Non-Commutative Polynomials." Annals of Mathematics, vol.34 (1933), pp.480-508.

73. Procesi, C. "A non-commutative Hilbert Nullstellensatz." Rendiconti di Mathematica, vol.25 (1966), pp.17-21.

74. ___________. "Non-Commutative Affine Rings." Atti Della Accademia Nazionale Dei Lincei, Memorie, vol.8 (1967), pp.232-255.

75. Rabin, M.O. "Diophantine Equations and Non-Standard Models of Arithmetic." Proceedings of the International Congress in Logic, Methodology, and Philosophy of Science, Stanford, 1960.

76. ___________. "Classes of Models and Sets of Sentences with the Intersection Property." Annals de la Faculté des Sciences de l'Universite de Clarmont, vol.1 (1962), pp.39-53.

77. Robinson, A. "Model Theory and Non-Standard Arithmetic." Infinitistic Methods, Proceedings of the Symposium on the Foundations of Mathematics, Warsaw, 1959.

78. ___________. "A Note on Embedding Problems." Fundamenta Mathematicae, vol.50 (1962), pp.445-461.

79. ___________. Introduction to Model Theory and to the Metamathematics of Algebra. Amsterdam, North-Holland Publishing Company, 1965.

80. ___________. "Forcing in Model Theory." Symposia Mathematica, vol.5 (1970), pp.64-82.

81. ___________. "Infinite Forcing in Model Theory." Proceedings of the Second Scandinavian Logic Symposium, Oslo, 1970. Amsterdam, North-Holland Publishing Company, 1971.

82. ___________. "Forcing in Model Theory." Actes des Congres International des Mathematicians, Nice, 1970.

83. ___________. "A Decision Method for Elementary Algebra and Geometry Revisited." To appear in the Proceedings of the Tarski Symposium, Berkeley, 1971.

84. ___________. "On the Notion of Algebraic Closedness for Noncommutative Groups and Fields." Journal of Symbolic Logic, vol.36 (1971), pp.441-444.

85. ___________. "Nonstandard Arithmetic and Generic Arithmetic." To appear in the Proceedings of the International Congress for Logic, Methodology, and Philosophy of Science, Bucharest, 1971.

86. Robinson, J. "Definability and Decision Problems in Arithmetic." Journal of Symbolic Logic, vol.14 (1949), pp.98-114.

87. Rogers, H. Theory of Recursive Functions and Effective Computability. New York, McGraw-Hill Book Company, 1967.

88. Sabbagh, G. "Sur la pureté dans les modules." C.R. Acad. Sc. Paris., vol.271 (1970), pp.865-867.

89. ___________. "Aspects Logiques de la pureté dans les modules." C.R. Acad. Sc. Paris, vol.271 (1970), pp.909-912.

90. ___________. "Sous-modules purs, existentiellement clos et élémentaires." C.R. Acad. Sc. Paris, vol.272 (1971), pp.1289-1292.

91. __________. "Embedding Problems for Modules and Rings with Application to Model-Companions." Journal of Algebra, vol.18 (1971), pp.390-403.

92. Sacks, G. "Higher Recursion Theory." Mimeographed notes.

93. Saracino, D. Notices of the American Mathematical Society, vol.18 (1971), p.668.

94. __________. Dissertation. Princeton University, 1972.

95. __________. "Model Companions for $\aleph_0$-Categorical Theories." Proceedings of the American Mathematical Society, vol.39 (1973), pp.591-598.

96. __________. "Wreath Products and Existentially Complete Solvable Groups." Transactions of the American Mathematical Society, vol.197 (1974), pp. 327-339.

97. __________. "A Counterexample in the Theory of Model-Companions." Preprint.

98. Saracino, D., and L. Lipshitz. "The model-companion of the theory of commutative rings without nilpotent elements." Proceedings of the American Mathematical Society, vol.38 (1973), pp.381-387.

99. Schenkman, E.V. "Roots of Centre Elements of Division Rings." Journal of the London Mathematical Society, vol.36 (1961), pp.393-398.

100. Scott, D. "Algebras of sets binumerable in complete extensions of arithmetic." Recursive Function Theory. (Proceedings of the Fifth Symposium in Pure Mathematics of the American Mathematical Society). Providence, American Mathematical Society, 1962.

101. Scott, W.R. "Algebraically Closed Groups." Proceedings of the American Mathematical Society, vol.2 (1951), pp.118-121.

102. Shelah, S. "A Note on Model Complete Models and Generic Models." Proceedings of the American Mathematical Society, vol.34 (1972), pp.509-514.

103. Shoenfield, J.R. Mathematical Logic. Reading, Massachusetts; Addison-Wesley Publishing Company; 1967.

104. Simmons, H. "Existentially Closed Structures." Journal of Symbolic Logic, vol.37 (1972), pp.293-310.

105. Smits, T.H.M. "Skew Polynomial Rings." Indagationes Mathematicae, vol.30 (1968), pp.209-224.

106. Szele, T. "Ein Analogen der Korpertheories für abelsche Gruppen." Journal für die reine und angewandte Mathematik, vol.188 (1950), pp.167-192.

107. Wheeler, W.H. Algebraically Closed Division Rings, Forcing, and the Analytical Hierarchy. Dissertation, Yale University, 1972.

108. Wood, C. Forcing for Infinitary Languages. Dissertation, Yale University, 1971.

109. Wood, C. "Forcing for Infinitary Languages." Zeitschrift für Mathematische Logik und Grundlagen der Mathematik, vol.18 (1972), pp.385-402.

110. Zariski, O. "A New Proof of Hilbert's Nullstellensatz." Bulletin of the American Mathematical Society, vol.53 (1947), pp.362-368.

111. Shelah, S. "Differentially Closed Fields." Israel Journal of Mathematics, vol.16 (1973), pp.314-328.

112. Macintyre, A. Abstract 74T-E38. Notices of the American Mathematical Society, vol.21 (1974), p.A379.

113. Macintyre, A., and D. Saracino. Abstract 74T-E37. Notices of the American Mathematical Society, vol.21 (1974), p.A379.

114. Saracino, D. Abstract 74T-E36. Notices of the American Mathematical Society, vol.21 (1974), p.A379.

115. __________. "m-existentially complete structures." Colloquium Mathematicum, vol.30 (1974), pp.7-13.

SUBJECT INDEX

Vol. 277: Séminaire Banach. Edité par C. Houzel. VII, 229 pages. 1972. DM 22,–

Vol. 278: H. Jacquet, Automorphic Forms on GL(2) Part II. XIII, 142 pages. 1972. DM 18,–

Vol. 279: R. Bott, S. Gitler and I. M. James, Lectures on Algebraic and Differential Topology. V, 174 pages. 1972. DM 20,–

Vol. 280: Conference on the Theory of Ordinary and Partial Differential Equations. Edited by W. N. Everitt and B. D. Sleeman. XV, 367 pages. 1972. DM 29,–

Vol. 281: Coherence in Categories. Edited by S. Mac Lane. VII, 235 pages. 1972. DM 22,–

Vol. 282: W. Klingenberg und P. Flaschel, Riemannsche Hilbertmannigfaltigkeiten. Periodische Geodätische. VII, 211 Seiten. 1972. DM 22,–

Vol. 283: L. Illusie, Complexe Cotangent et Déformations II. VII, 304 pages. 1972. DM 27,–

Vol. 284: P. A. Meyer, Martingales and Stochastic Integrals I. VI, 89 pages. 1972. DM 18,–

Vol. 285: P. de la Harpe, Classical Banach-Lie Algebras and Banach-Lie Groups of Operators in Hilbert Space. III, 160 pages. 1972. DM 18,–

/ol. 286: S. Murakami, On Automorphisms of Siegel Domains. V, 95 pages. 1972. DM 18,–

/ol. 287: Hyperfunctions and Pseudo-Differential Equations. Edited by H. Komatsu. VII, 529 pages. 1973. DM 40,–

/ol. 288: Groupes de Monodromie en Géométrie Algébrique. (SGA 7 I). Dirigé par A. Grothendieck. IX, 523 pages. 1972. DM 55,–

/ol. 289: B. Fuglede, Finely Harmonic Functions. III, 188. 1972. DM 20,–

/ol. 290: D. B. Zagier, Equivariant Pontrjagin Classes and Applications to Orbit Spaces. IX, 130 pages. 1972. DM 18,–

/ol. 291: P. Orlik, Seifert Manifolds. VIII, 155 pages. 1972. DM 18,–

'ol. 292: W. D. Wallis, A. P. Street and J. S. Wallis, Combinatorics: Room Squares, Sum-Free Sets, Hadamard Matrices. V, 508 pages. 1972. DM 55,–

'ol. 293: R. A. DeVore, The Approximation of Continuous Functions by Positive Linear Operators. VIII, 289 pages. 1972. DM 27,–

'ol. 294: Stability of Stochastic Dynamical Systems. Edited by R. F. Curtain. IX, 332 pages. 1972. DM 29,–

'ol. 295: C. Dellacherie, Ensembles Analytiques Capacités. Mesures de Hausdorff. XII, 123 pages. 1972. DM 18,–

'ol. 296: Probability and Information Theory II. Edited by M. Behara, K. Krickeberg and J. Wolfowitz. V, 223 pages. 1973. DM 22,–

'ol. 297: J. Garnett, Analytic Capacity and Measure. IV, 138 pages. 1972. DM 18,–

ol. 298: Proceedings of the Second Conference on Compact Transformation Groups. Part 1. XIII, 453 pages. 1972. DM 35,–

'ol. 299: Proceedings of the Second Conference on Compact Transformation Groups. Part 2. XIV, 327 pages. 1972. DM 29,–

ol. 300: P. Eymard, Moyennes Invariantes et Représentations Unitaires. II, 113 pages. 1972. DM 18,–

ol. 301: F. Pittnauer, Vorlesungen über asymptotische Reihen. VI, 186 Seiten. 1972. DM 18,–

ol. 302: M. Demazure, Lectures on p-Divisible Groups. V, 98 pages. 1972. DM 18,–

ol. 303: Graph Theory and Applications. Edited by Y. Alavi, D. R. Lick and A. T. White. IX, 329 pages. 1972. DM 26,–

ol. 304: A. K. Bousfield and D. M. Kan, Homotopy Limits, Completions and Localizations. V, 348 pages. 1972. DM 29,–

ol. 305: Théorie des Topos et Cohomologie Etale des Schémas. Tome 3. (SGA 4). Dirigé par M. Artin, A. Grothendieck et J. L. Verdier. I, 640 pages. 1973. DM 55,–

ol. 306: H. Luckhardt, Extensional Gödel Functional Interpretation. I, 161 pages. 1973. DM 20,–

ol. 307: J. L. Bretagnolle, S. D. Chatterji et P.-A. Meyer, Ecole d'été de Probabilités: Processus Stochastiques. VI, 198 pages. 1973. DM 22,–

ol. 308: D. Knutson, λ-Rings and the Representation Theory of the Symmetric Group. IV, 203 pages. 1973. DM 22,–

Vol. 309: D. H. Sattinger, Topics in Stability and Bifurcation Theory. VI, 190 pages. 1973. DM 20,–

Vol. 310: B. Iversen, Generic Local Structure of the Morphisms in Commutative Algebra. IV, 108 pages. 1973. DM 18,–

Vol. 311: Conference on Commutative Algebra. Edited by J. W. Brewer and E. A. Rutter. VII, 251 pages. 1973. DM 24,–

Vol. 312: Symposium on Ordinary Differential Equations. Edited by W. A. Harris, Jr. and Y. Sibuya. VIII, 204 pages. 1973. DM 22,–

Vol. 313: K. Jörgens and J. Weidmann, Spectral Properties of Hamiltonian Operators. III, 140 pages. 1973. DM 18,–

Vol. 314: M. Deuring, Lectures on the Theory of Algebraic Functions of One Variable. VI, 151 pages. 1973. DM 18,–

Vol. 315: K. Bichteler, Integration Theory (with Special Attention to Vector Measures). VI, 357 pages. 1973. DM 29,–

Vol. 316: Symposium on Non-Well-Posed Problems and Logarithmic Convexity. Edited by R. J. Knops. V, 176 pages. 1973. DM 20,–

Vol. 317: Séminaire Bourbaki – vol. 1971/72. Exposés 400–417. IV, 361 pages. 1973. DM 29,–

Vol. 318: Recent Advances in Topological Dynamics. Edited by A. Beck. VIII, 285 pages. 1973. DM 27,–

Vol. 319: Conference on Group Theory. Edited by R. W. Gatterdam and K. W. Weston. V, 188 pages. 1973. DM 20,–

Vol. 320: Modular Functions of One Variable I. Edited by W. Kuyk. V, 195 pages. 1973. DM 20,–

Vol. 321: Séminaire de Probabilités VII. Edité par P. A. Meyer. VI, 322 pages. 1973. DM 29,–

Vol. 322: Nonlinear Problems in the Physical Sciences and Biology. Edited by I. Stakgold, D. D. Joseph and D. H. Sattinger. VIII, 357 pages. 1973. DM 29,–

Vol. 323: J. L. Lions, Perturbations Singulières dans les Problèmes aux Limites et en Contrôle Optimal. XII, 645 pages. 1973. DM 46,–

Vol. 324: K. Kreith, Oscillation Theory. VI, 109 pages. 1973. DM 18,–

Vol. 325: C.-C. Chou, La Transformation de Fourier Complexe et L'Equation de Convolution. IX, 137 pages. 1973. DM 18,–

Vol. 326: A. Robert, Elliptic Curves. VIII, 264 pages. 1973. DM 24,–

Vol. 327: E. Matlis, One-Dimensional Cohen-Macaulay Rings. XII, 157 pages. 1973. DM 20,–

Vol. 328: J. R. Büchi and D. Siefkes, The Monadic Second Order Theory of All Countable Ordinals. VI, 217 pages. 1973. DM 22,–

Vol. 329: W. Trebels, Multipliers for (C, α)-Bounded Fourier Expansions in Banach Spaces and Approximation Theory. VII, 103 pages. 1973. DM 18,–

Vol. 330: Proceedings of the Second Japan-USSR Symposium on Probability Theory. Edited by G. Maruyama and Yu. V. Prokhorov. VI, 550 pages. 1973. DM 40,–

Vol. 331: Summer School on Topological Vector Spaces. Edited by L. Waelbroeck. VI, 226 pages. 1973. DM 22,–

Vol. 332: Séminaire Pierre Lelong (Analyse) Année 1971-1972. V, 131 pages. 1973. DM 18,–

Vol. 333: Numerische, insbesondere approximationstheoretische Behandlung von Funktionalgleichungen. Herausgegeben von R. Ansorge und W. Törnig. VI, 296 Seiten. 1973. DM 27,–

Vol. 334: F. Schweiger, The Metrical Theory of Jacobi-Perron Algorithm. V, 111 pages. 1973. DM 18,–

Vol. 335: H. Huck, R. Roitzsch, U. Simon, W. Vortisch, R. Walden, B. Wegner und W. Wendland, Beweismethoden der Differentialgeometrie im Großen. IX, 159 Seiten. 1973. DM 20,–

Vol. 336: L'Analyse Harmonique dans le Domaine Complexe. Edité par E. J. Akutowicz. VIII, 169 pages. 1973. DM 20,–

Vol. 337: Cambridge Summer School in Mathematical Logic. Edited by A. R. D. Mathias and H. Rogers. IX, 660 pages. 1973. DM 46,–

Vol: 338: J. Lindenstrauss and L. Tzafriri, Classical Banach Spaces. IX, 243 pages. 1973. DM 24,–

Vol. 339: G. Kempf, F. Knudsen, D. Mumford and B. Saint-Donat, Toroidal Embeddings I. VIII, 209 pages. 1973. DM 22,–

Vol. 340: Groupes de Monodromie en Géométrie Algébrique. (SGA 7 II). Par P. Deligne et N. Katz. X, 438 pages. 1973. DM 44,–

Vol. 341: Algebraic K-Theory I, Higher K-Theories. Edited by H. Bass. XV, 335 pages. 1973. DM 29,–

Vol. 342: Algebraic K-Theory II, "Classical" Algebraic K-Theory, and Connections with Arithmetic. Edited by H. Bass. XV, 527 pages. 1973. DM 40,-

Vol. 343: Algebraic K-Theory III, Hermitian K-Theory and Geometric Applications. Edited by H. Bass. XV, 572 pages. 1973. DM 40,-

Vol. 344: A. S. Troelstra (Editor), Metamathematical Investigation of Intuitionistic Arithmetic and Analysis. XVII, 485 pages. 1973. DM 38,-

Vol. 345: Proceedings of a Conference on Operator Theory. Edited by P. A. Fillmore. VI, 228 pages. 1973. DM 22,-

Vol. 346: Fučík et al., Spectral Analysis of Nonlinear Operators. II, 287 pages. 1973. DM 26,-

Vol. 347: J. M. Boardman and R. M. Vogt, Homotopy Invariant Algebraic Structures on Topological Spaces. X, 257 pages. 1973. DM 24,-

Vol. 348: A. M. Mathai and R. K. Saxena, Generalized Hypergeometric Functions with Applications in Statistics and Physical Sciences. VII, 314 pages. 1973. DM 26,-

Vol. 349: Modular Functions of One Variable II. Edited by W. Kuyk and P. Deligne. V, 598 pages. 1973. DM 38,-

Vol. 350: Modular Functions of One Variable III. Edited by W. Kuyk and J.-P. Serre. V, 350 pages. 1973. DM 26,-

Vol. 351: H. Tachikawa, Quasi-Frobenius Rings and Generalizations. XI, 172 pages. 1973. DM 20,-

Vol. 352: J. D. Fay, Theta Functions on Riemann Surfaces. V, 137 pages. 1973. DM 18,-

Vol. 353: Proceedings of the Conference on Orders, Group Rings and Related Topics. Organized by J. S. Hsia, M. L. Madan and T. G. Ralley. X, 224 pages. 1973. DM 22,-

Vol. 354: K. J. Devlin, Aspects of Constructibility. XII, 240 pages. 1973. DM 24,-

Vol. 355: M. Sion, A Theory of Semigroup Valued Measures. V, 140 pages. 1973. DM 18,-

Vol. 356: W. L. J. van der Kallen, Infinitesimally Central Extensions of Chevalley Groups. VII, 147 pages. 1973. DM 18,-

Vol. 357: W. Borho, P. Gabriel und R. Rentschler, Primideale in Einhüllenden auflösbarer Lie-Algebren. V, 182 Seiten. 1973. DM 20,-

Vol. 358: F. L. Williams, Tensor Products of Principal Series Representations. VI, 132 pages. 1973. DM 18,-

Vol. 359: U. Stammbach, Homology in Group Theory. VIII, 183 pages. 1973. DM 20,-

Vol. 360: W. J. Padgett and R. L. Taylor, Laws of Large Numbers for Normed Linear Spaces and Certain Fréchet Spaces. VI, 111 pages. 1973. DM 18,-

Vol. 361: J. W. Schutz, Foundations of Special Relativity: Kinematic Axioms for Minkowski Space Time. XX, 314 pages. 1973. DM 26,-

Vol. 362: Proceedings of the Conference on Numerical Solution of Ordinary Differential Equations. Edited by D. Bettis. VIII, 490 pages. 1974. DM 34,-

Vol. 363: Conference on the Numerical Solution of Differential Equations. Edited by G. A. Watson. IX, 221 pages. 1974. DM 20,-

Vol. 364: Proceedings on Infinite Dimensional Holomorphy. Edited by T. L. Hayden and T. J. Suffridge. VII, 212 pages. 1974. DM 20,-

Vol. 365: R. P. Gilbert, Constructive Methods for Elliptic Equations. VII, 397 pages. 1974. DM 26,-

Vol. 366: R. Steinberg, Conjugacy Classes in Algebraic Groups (Notes by V. V. Deodhar). VI, 159 pages. 1974. DM 18,-

Vol. 367: K. Langmann und W. Lütkebohmert, Cousinverteilungen und Fortsetzungssätze. VI, 151 Seiten. 1974. DM 16,-

Vol. 368: R. J. Milgram, Unstable Homotopy from the Stable Point of View. V, 109 pages. 1974. DM 16,-

Vol. 369: Victoria Symposium on Nonstandard Analysis. Edited by A. Hurd and P. Loeb. XVIII, 339 pages. 1974. DM 26,-

Vol. 370: B. Mazur and W. Messing, Universal Extensions and One Dimensional Crystalline Cohomology. VII, 134 pages. 1974. DM 16,-

Vol. 371: V. Poenaru, Analyse Différentielle. V, 228 pages. 1974. DM 20,-

Vol. 372: Proceedings of the Second International Conference on the Theory of Groups 1973. Edited by M. F. Newman. VII, 740 pages. 1974. DM 48,-

Vol. 373: A. E. R. Woodcock and T. Poston, A Geometrical Study of the Elementary Catastrophes. V, 257 pages. 1974. DM 22,-

Vol. 374: S. Yamamuro, Differential Calculus in Topological Linear Spaces. IV, 179 pages. 1974. DM 18,-

Vol. 375: Topology Conference 1973. Edited by R. F. Dickman Jr. and P. Fletcher. X, 283 pages. 1974. DM 24,-

Vol. 376: D. B. Osteyee and I. J. Good, Information, Weight of Evidence, the Singularity between Probability Measures and Signal Detection. XI, 156 pages. 1974. DM 16,-

Vol. 377: A. M. Fink, Almost Periodic Differential Equations. VIII, 336 pages. 1974. DM 26,-

Vol. 378: TOPO 72 - General Topology and its Applications. Proceedings 1972. Edited by R. Alò, R. W. Heath and J. Nagata. XIV, 651 pages. 1974. DM 50,-

Vol. 379: A. Badrikian et S. Chevet, Mesures Cylindriques, Espaces de Wiener et Fonctions Aléatoires Gaussiennes. X, 383 pages. 1974. DM 32,-

Vol. 380: M. Petrich, Rings and Semigroups. VIII, 182 pages. 1974. DM 18,-

Vol. 381: Séminaire de Probabilités VIII. Edité par P. A. Meyer. IX, 354 pages. 1974. DM 32,-

Vol. 382: J. H. van Lint, Combinatorial Theory Seminar Eindhoven University of Technology. VI, 131 pages. 1974. DM 18,-

Vol. 383: Séminaire Bourbaki - vol. 1972/73. Exposés 418-435 IV, 334 pages. 1974. DM 30,-

Vol. 384: Functional Analysis and Applications, Proceedings 1972. Edited by L. Nachbin. V, 270 pages. 1974. DM 22,-

Vol. 385: J. Douglas Jr. and T. Dupont, Collocation Methods for Parabolic Equations in a Single Space Variable (Based on C^1-Piecewise-Polynomial Spaces). V, 147 pages. 1974. DM 16,-

Vol. 386: J. Tits, Buildings of Spherical Type and Finite BN-Pairs. IX, 299 pages. 1974. DM 24,-

Vol. 387: C. P. Bruter, Eléments de la Théorie des Matroïdes. V, 138 pages. 1974. DM 18,-

Vol. 388: R. L. Lipsman, Group Representations. X, 166 pages. 1974. DM 20,-

Vol. 389: M.-A. Knus et M. Ojanguren, Théorie de la Descente et Algèbres d' Azumaya. IV, 163 pages. 1974. DM 20,-

Vol. 390: P. A. Meyer, P. Priouret et F. Spitzer, Ecole d'Eté de Probabilités de Saint-Flour III - 1973. Edité par A. Badrikian et P.-L. Hennequin. VIII, 189 pages. 1974. DM 20,-

Vol. 391: J. Gray, Formal Category Theory: Adjointness for 2-Categories. XII, 282 pages. 1974. DM 24,-

Vol. 392: Géométrie Différentielle, Colloque, Santiago de Compostela, Espagne 1972. Edité par E. Vidal. VI, 225 pages. 1974. DM 20,-

Vol. 393: G. Wassermann, Stability of Unfoldings. IX, 164 pages. 1974. DM 20,-

Vol. 394: W. M. Patterson 3rd, Iterative Methods for the Solution of a Linear Operator Equation in Hilbert Space - A Survey. III, 183 pages. 1974. DM 20,-

Vol. 395: Numerische Behandlung nichtlinearer Integrodifferential- und Differentialgleichungen. Tagung 1973. Herausgegeben von R. Ansorge und W. Törnig. VII, 313 Seiten. 1974. DM 28,-

Vol. 396: K. H. Hofmann, M. Mislove and A. Stralka, The Pontryagin Duality of Compact O-Dimensional Semilattices and its Applications. XVI, 122 pages. 1974. DM 18,-

Vol. 397: T. Yamada, The Schur Subgroup of the Brauer Group. V, 159 pages. 1974. DM 18,-

Vol. 398: Théories de l'Information, Actes des Rencontres de Marseille-Luminy, 1973. Edité par J. Kampé de Fériet et C. Picard. XII, 201 pages. 1974. DM 23,-